AF711610

COGNITIVE REHABILITATION OF MEMORY

A Clinical-Neuropsychological Introduction

COGNITIVE REHABILITATION OF MEMORY

A Clinical-Neuropsychological Introduction

HELMUT HILDEBRANDT

ELSEVIER

ACADEMIC PRESS

An imprint of Elsevier

Academic Press is an imprint of Elsevier
125 London Wall, London EC2Y 5AS, United Kingdom
525 B Street, Suite 1650, San Diego, CA 92101, United States
50 Hampshire Street, 5th Floor, Cambridge, MA 02139, United States
The Boulevard, Langford Lane, Kidlington, Oxford OX5 1GB, United Kingdom

© 2019 Elsevier Inc. All rights reserved.

No part of this publication may be reproduced or transmitted in any form or by any means, electronic or mechanical, including photocopying, recording, or any information storage and retrieval system, without permission in writing from the publisher. Details on how to seek permission, further information about the Publisher's permissions policies and our arrangements with organizations such as the Copyright Clearance Center and the Copyright Licensing Agency, can be found at our website: www.elsevier.com/permissions.

This book and the individual contributions contained in it are protected under copyright by the Publisher (other than as may be noted herein).

Notices
Knowledge and best practice in this field are constantly changing. As new research and experience broaden our understanding, changes in research methods, professional practices, or medical treatment may become necessary.

Practitioners and researchers must always rely on their own experience and knowledge in evaluating and using any information, methods, compounds, or experiments described herein. In using such information or methods they should be mindful of their own safety and the safety of others, including parties for whom they have a professional responsibility.

To the fullest extent of the law, neither the Publisher nor the authors, contributors, or editors, assume any liability for any injury and/or damage to persons or property as a matter of products liability, negligence or otherwise, or from any use or operation of any methods, products, instructions, or ideas contained in the material herein.

Library of Congress Cataloging-in-Publication Data
A catalog record for this book is available from the Library of Congress

British Library Cataloguing-in-Publication Data
A catalogue record for this book is available from the British Library

ISBN 978-0-12-816981-0

For information on all Academic Press publications visit our website at https://www.elsevier.com/books-and-journals

Working together to grow libraries in developing countries

www.elsevier.com • www.bookaid.org

Publisher: Nikki Levy
Acquisition Editor: Joslyn Chaiprasert-Paguio
Editorial Project Manager: Tracy Tufaga
Production Project Manager: Punithavathy Govindaradjane
Cover Designer: Matthew Limbert

Typeset by SPi Global, India

DEDICATION

To all the patients I treated and to my colleagues, which were involved in the research: Claudia Mödden, PhD, Barbar Mussmann-Mork, PhD, and Merle Richter.

CONTENTS

FOREWORD

Helmut Hildebrandt has practiced as a clinical neuropsychologist for several decades. In the neurology department in Bremen he had ample opportunity to observe and treat patients with a wide variety of neurological disorders. Moreover, he has been involved in many research projects, dealing with fundamental neuropsychological issues, assessment questions, and cognitive treatment trials. He has coauthored a very large number of scientific papers, in particular when one takes his clinical activities into account. He has now written a book, based on his experiences, insights, and findings on the do's and don'ts for memory rehabilitation. His views are clearly based on theoretical insights and empirical findings. Moreover, he is sensitive to the many methodological hurdles associated with rehabilitation research. All of this has resulted in a book that lays out his views that in some respects differ from the "traditional approach" developing at the end of the 20th century.

In the 1960s and 1970s, neuropsychologists were primarily interested in aphasia and laterality. In the 1980s a shift toward memory-related issues can be observed. The English cognitive psychologist Allan Baddeley introduced, with Graham Hitch, the Working Memory model in 1974. He also began to collaborate with colleagues on clinical neuropsychological issues and this combination of theory-driven and applied clinical studies has proven very fruitful. This line of research produced new insights in the nature of cognitive disorders in patients and resulted also in new neuropsychological instruments.

Another major step forwards was made by Barbara Wilson, who wrote her dissertation under supervision of Baddeley and she published her findings in her book *Rehabilitation of Memory* in 1988. That book formed a major impulse in the area of cognitive rehabilitation. Rehabilitation of language disorders had been around for decades but was the territory of speech therapists. Rehabilitation in other areas of cognitive disorders was only beginning to develop and often was limited to skills training.

It became clear that standard computer programs with simple memory tasks were not going to be effective for most patients, in particular for patients with more severe problems. The conclusion of the first period of memory rehabilitation studies was that memory was not like a muscle and would not profit from repetitive practice. An alternative option would be to focus on the needs of an individual patient and develop compensatory strategies with which some crucial information could be retained and recalled. Sometimes electronic devices were used to compensate for lost mental capacities.

Hildebrandt argues in his book that the single-case approach that evolved from this development may be useful for patients with severe memory disorders. However, there are large numbers of patients that suffer from mild to moderate memory problems for which a treatment should be found that is both profitable to them and feasible in the clinic. In order to present his case he first digs deep into methodological problems associated with perhaps the most difficult branch of clinical research, the evaluation of a therapy. There are many confounding factors that may produce a seemingly positive effect that in fact should not be attributed to the critical elements of a treatment. A therapy may have a short-term effect, but what will be the effect after, say, one year? The majority of cognitive rehabilitation studies fail to measure such long-term effects. Perhaps the most crucial problem is that of generalizabilty or transfer to daily life. Normally, treatment is offered in a clinical setting but the patient has to use the newly acquired skills in his daily life. Of course, one can ask an individual whether he thinks he has improved and whether he has less complaints in his daily life. But producing observational data of such an improvement is something different.

Hildebrandt discusses the many methodological and statistical issues associated with performing rehabilitation studies in detail. He illustrates the issues by elaborating on examples from the literature. By doing so, he presents the blueprints of protocols for treatment studies that are strong enough to provide significant results, not only from a statistical perspective but also from a clinical perspective.

Perhaps more importantly, Hildebrandt argues that we need to develop more cleverly designed treatments. Memory researchers have discovered many factors that may enhance memory recall in healthy individuals and these factors relate to how information is encoded, what happens during the consolidation phase, and how information has to be retrieved during recall or recognition. These insights may help to develop treatment protocols that are fruitful in the clinic, in particular for patients with mild to moderate memory problems. In recent years, Hildebrandt and his colleagues have developed some protocols that are based on such insights. For instance, they developed a protocol focusing on semantic structuring of incoming information. In another study they looked at the repetition lag procedure. Such phenomena are not surprising to the fundamental memory researcher, but they have now been applied in an innovative way in memory treatments and with success.

Neuropsychology as a scientific discipline has revealed many remarkable features of how the brain functions and produces behavior. However, most neuropsychologists deal with patients with neurological disorders and a very large proportion also suffers from memory problems, of varying degrees. Objectifying these memory disorders is important, but providing suggestions or exercises helping patients how to deal with these problems and even increase memory performance is of course of much more value for these patients. I hope Hildebrandt's book will stimulate many neuropsychologists to get to work to focus on proper ways of providing these treatments and evaluating their value with methodologically sound studies.

Paul Eling

PREFACE

Memory disorders are frequent consequences of neurological diseases. In this respect, one may expect that their treatment is based on a deep scientific understanding of memory and that there are explicit guidelines for helping memory-impaired patients regain their functional abilities. Both are not the case. Recent Cochrane reviews on memory rehabilitation argue that there is no convincing evidence of a positive effect for either pharmacologically or functionally oriented interventions for stroke patients (das Nair, Cogger, Worthington, & Lincoln, 2016, 2017). There is also no evidence for a pharmacological treatment of memory disorders in multiple sclerosis (He et al., 2013) or in traumatic brain injury (Dougall, Poole, & Agrawal, 2015). An impact of Acetylcholinesterase Inhibitors drugs in dementia is present but limited (Evans, Wilcock, & Birks, 2004). On the other hand, the recent German S3 guideline for the treatment of dementia denies any effect of a neuropsychological treatment in demented patients (Bahar-Fuchs, Clare, & Woods, 2013). There is some evidence for an effective neuropsychological treatment of memory disorders in patients with traumatic brain injury (Mahan, Rous, & Adlam, 2017) and in multiple sclerosis (das Nair, Martin, & Lincoln, 2016), but to date this evidence is based upon the results of only a few studies. To name just one other phenomenon showing that research on memory rehabilitation is still in its infancy: The last book with a similar title was published in 2009, and most of the textbooks on this topic are written by just one author, i.e. Barbara A. Wilson.

Clearly, there have been many reasons why research on memory rehabilitation has not developed as rapidly as in other sections of neurorehabilitation (if you compare, for example, the growing number of publications on motor or neglect rehabilitation). In my view, the following causes play at least some role. Research on memory rehabilitation started with two decisions, both of which were unfortunate for its further development. First, B. Wilson (1987) published a very influential paper in 1987 arguing to use single-case experimental design instead of group studies in rehabilitation research. At the same time, medicine became increasingly oriented towards rigorously defined, randomized controlled trials ("evidence-based medicine"). Consequently, results from single-case experimental design studies eventually occupied an outsider position and lost credit. Second, and even more important, due to the national specificity of the health system in the

United Kingdom, Wilson started to focus her research on chronic and severely memory-impaired patients. Unfortunately, for this group of patients the prospect for substantial recovery is very low. Therefore, Wilson stated that a remission of memory disorder is more or less impossible (B. A. Wilson, 1987). Stressing the importance of behavioral modification approaches and the use of external memory aids for rehabilitation was a natural consequence of working with such patients. As a logical side effect of a focus on severely impaired patients and on a function that shows no recovery, cognitive models of memory lost their meaning for neuropsychological rehabilitation attempts. This is not to say that Wilson was not interested in cognitive models of memory – the opposite was the case. This is documented by the number of distinctive articles she wrote or coauthored on this topic. While cognitive models of memory and memory disorder were useful for diagnosis, they did not play an essential role in treating the type of memory-impaired patients that she was working with.

Because many researchers followed in the steps of Barbara Wilson during the two decades between 1990 and 2010, research on memory rehabilitation lost its foundation in cognitive models of memory. At the same time, the development of functional MRI shifted the interest of neuroscientists and university-based neuropsychologists strongly in the direction of neuroanatomical, localizationist, or simple technical questions; ignoring the need to develop new cognitive models for learning and memory that could be used to help memory-impaired patients. The very productive interaction between university-based research, clinical neuropsychology, and neuropsychological rehabilitation (which had pushed the field forward during the decades before the occurrence of fMRI) vanished almost completely. As far as I can see, in the years since the spread of fMRI at the universities no major translational impact has been generated from the neuroscience of memory for memory rehabilitation. Furthermore, the results of memory rehabilitation attempts have not been useful for the development of new models of (impaired) memory. Potential new models could have influenced fundamental science research at universities, so their absence is noteworthy.

This diagnosis of the state of memory rehabilitation may sound slightly too skeptical. In any case, it was the driving background for compiling this book and it structured the content and sequence of its chapters. Having argued above that the decision to focus on single case studies was strategically wrong, this book starts with a section on evidence-based medicine, on different evidence levels, and on what randomized controlled trials imply, as well as why they are so important for our treatment knowledge. Delivering well-designed randomized controlled trials on cognitive (and memory) rehabilitation will have a decisive impact on the question of whether clinical neuropsychology remains a central aspect of neurorehabilitation in the future. Measuring the outcome of memory research on a level that is appropriate for scientific progress, as well as increasing the functional independence of patients are two core elements. At the same time, I will argue that single case experimental design studies should play *some* role in neuropsychological rehabilitation research. For example, these studies would be beneficial when trying out new treatment ideas or the potential number of patients with a specific syndrome is too small to focus on group studies.

The second major starting point of this book is to focus on patients with mild to moderate memory impairment. It is a common wisdom in the field of neurorehabilitation that functional recovery in a strong sense presupposes that the impaired function still can be executed by the patient to some degree. For example, in constraint-induced movement therapy (a method with a high amount of evidence in motor rehabilitation), one major exclusion criterion is if the patient suffers from a complete paresis. In the event of a complete paresis (which lasts for multiple weeks), the prospect of any recovery is very low to absent. A short review of the existing animal studies on memory rehabilitation (which is part of this book), shows that as in motor paresis, recovery from memory deficits is possible. However, this depends on the severity of the impairment and the localization of the lesion: Severe and focal memory impairments that last for a longer time period show little prospect of recovery. This is a quite similar finding to that which is found in motor rehabilitation. Therefore, it seems wise to distinguish between different goals and strategies in the rehabilitation of mild to moderately impaired patients compared to severely impaired patients.

Focusing on moderate and mild memory impairments with some prospects for recovery, a third starting point of this book is to argue that the development of interventions for memory rehabilitation must be based on cognitive models of memory. Only such models will help research to focus on the specific subfunctions and processes that are impaired and that can be targeted by specific trainings. A background in cognitive models is also a presupposition to initiate further translational interaction with university research, focusing on neuroscientific and technical questions. There is yet another reason to underline the central meaning of cognitive models for memory rehabilitation: only such models allow for the decision regarding what type of intervention may plausibly target memory relevant processing and what kind of intervention does not. There are many existing interventions that assume positive effects on memory performance (for example neurofeedback of sensorimotor rhythms) without being able to offer an intelligible explanation for that impact. Therefore, it seems to be necessary to remember that medical interventions always presuppose at least some biological understanding of why specific drugs are helpful to the patients.

For this book, I decided to use the Atkinson and Shiffrin (1968) model as my starting point to review research on memory rehabilitation from the last 40 years. Clearly, this model is very outdated in many respects. Its use was motivated by two different reasons. First, because cognitive models have not played any major role in memory rehabilitation since the 1990s, the Atkinson and Shiffrin (1968) model pragmatically still covered many attempts in the field. Second, the book is subdivided into quasi-historical steps, starting with the earliest and more simplistic idea to use intact mnemonics to help memory-impaired patients and moving forward to more complicated models for memory and memory rehabilitation. In other words, from using intact encoding functions to improving encoding itself as a possible role for impaired memory; from teaching strategies at encoding to teaching transfer appropriate processing; from focusing on encoding to focusing on consolidation, and on retrieval for improving memory performance in impaired patients. According to these developments and with some reference to memory research conducted at universities, I tried to adapt the old Atkinson and Shiffrin (1968) model step by step and to explain "where" the interventions for improving memory might have worked (i.e. what kind of memory model we have to assume for the treatment attempts). Atkinson and Shiffrin (1968) postulate three different modules (or at least, *functional stages*) for memory: encoding, storage, and retrieval. Sorting the different strategies of memory rehabilitation along these stages shows that most of them have focused on improving the encoding. Meanwhile, there are also some serious attempts to improve retrieval and consolidation of memory in memory-impaired patients. Being explicit about which stage the intended treatment should work and being aware of the other steps of memory performance might assist in developing new and more specific ideas for functional interventions.

As I have worked for over two decades with patients, it is clear to me that there is a small but important group with profound and chronic memory impairments for whom such an approach as mentioned above would fail and contradict their needs. Therefore, in the last chapter of this book, evidence for severely-impaired patients and for teaching memory aids is reviewed. Of note is that I tried to derive some practical guidelines for colleagues working at clinics, which can be found at the end of both major chapters on memory rehabilitation in this book. This also pertains to the question of what type of memory aid should be used in different patients and what type of learning strategy might be attempted depending on the severity of the impairment. Unfortunately, the possibility for a specific guideline of interventions is very limited – due to the limited number of well-designed studies in the field. Due to this insight, the major intention of writing this book was to inspire colleagues to think about memory rehabilitation in terms of cognitive models. It is my intention that readers will be able to use these models to take steps towards understanding which cognitive processes can be treated with prospects for a partial recovery of the impairment and which processes cannot.

CHAPTER I

Evidence-Based Medicine in Rehabilitation Research

1 INTRODUCTION

Neuropsychological rehabilitation is a part of neurological rehabilitation; the discipline of helping patients with acquired brain damage cope with the short- and long-term consequences of their disease. The definition and goals of neuropsychological rehabilitation are therefore included in a definition of neurological rehabilitation. The question thus becomes: How is neurological rehabilitation defined? On a very global level, one can define it as successful attempt to improve independence in daily life, the ability to work, and the ability to attain satisfying interpersonal relationships in patients that suffer from enduring consequences of organic brain damage. Being focused on the goal of neurological rehabilitation, this definition is not specific enough to distinguish it from other medical and health interventions. For example, treating patients with antiepileptic drugs also has the goal of helping patients cope with the consequences of organic brain damage and to increase their independence. Therefore, the definition above must be complemented by a specification of the methods of neurological rehabilitation that clearly differ from other attempts in health care. In contrast to medical interventions, neurological rehabilitation uses specific training schedules, introduction of external aids, and environmental restructuring as treatments. Pharmacological, surgical, and injection-based methods are used in rehabilitation, but only as an additional, accompanying, and endorsing treatment (not as an essential method). Rehabilitation may be defined as a learning process of lost functions, or as an external and internal adaptation to a temporary or chronic impairment due to damage of the nervous system.

Two aspects of this definition of neurorehabilitation should be discussed a little bit further before moving on to its more specific goal settings. First, neurorehabilitation shares a common mechanism with psychotherapy, i.e., learning. One may argue that the difference between behavioral therapy and neurorehabilitation is the etiology of maladaptation: Classical behavioral therapy concerns functionally acquired maladaptive behavior, or behavior

Cognitive Rehabilitation of Memory
https://doi.org/10.1016/B978-0-12-816981-0.00001-1
© 2019 Elsevier Inc.
All rights reserved.

that relies on problematic reinforcing schedules of the environment, or on problematic cognitive frames for interpersonal relationships. Neurorehabilitation concerns somatically acquire maladaptive behavior due to the loss of a motor, sensory, or cognitive ability.

Although there is clearly a parallel between behavioral therapy and neurorehabilitation, there are also some fundamental differences. The first difference is that most of the impairments do not concern the *behavioral* intentions of the patients, but rather their poor performance in specific cognitive processes. The treatment is focused to enhance and speed up relearning. The correction of maladaptive behavior is only a secondary aspect, which comes into play if the patients choose solutions that are not optimal or even limiting. The second difference is that patients in neurorehabilitation do not feel like patients in psychotherapy: They suffer from an often acute and significant disease that causes the (often focal) impairment. They must live with that impairment and try to improve their functioning without a long period of experience with it and without any clear prognosis of its development. In psychotherapy, the patients suffer due to a combination of internal and environmental factors, but this suffering often has a long history. Patients with neuropsychological impairment may also despair and need psychotherapy, but this is of comparatively smaller concern and is more often present or urgent during the chronic phase, i.e., at a time point where no further recovery of the impaired functions are expected.

The second aspect of the definition of neurorehabilitation that should be made more explicit is that attempting to achieve a maximum level of independence for the patient through neurorehabilitation may lead to a profound conflict between the professionals and the patients. Many patients find solutions for adapting to their impairment and handicap that are suboptimal. These solutions may rely on the help offered by relatives or nurses. They may also imply a restriction in mobility or in working abilities (early retirement, etc.). As defined above, neurorehabilitation is not focused directly on the quality of life of the patients, but instead on their maximal independence. One may argue that attaining the highest degree of independence is a prerequisite to reach the highest quality of life. Obviously, that would be a philosophical statement, and not always true for every patient. In the process of offering neurorehabilitation, it is often the case that patients are exhausted and refrain from additional efforts in training, despite the assertion of therapists that additional progress can be made. The opposite may also be the case: Patients might ask for more and specific training sessions without making further progress. Following the definition for neurorehabilitation as provided above will lead to actions that may "harm" the quality of life of the patients. In this respect, neurorehabilitation does not differ from other medical interventions: The goal is more to help the patients live independently and less to increase their happiness (but each therapist would be happy to do so. As it relies on learning processes, neurorehabilitation is only possible in cooperation with the patient). Consequently, goal setting and timing of rehabilitative effort must be *legitimized*, because it is the patient and his or her relatives that must be convinced of the reasons leading to the ultimate decision.

2 STRATEGIES OF NEUROREHABILITATION: RESTITUTION AND COMPENSATION

The ultimate goal of neurorehabilitation is the recovery of the patient in all his or her abilities and the attainment of a high quality of life. This should happen as cost effectively as possible. In scientific terms this goal is called restitution. Restitution after acquired brain damage implies either the possibility of regeneration of nervous tissue or the assumption of neuronal reserves after sustained brain damage. Both implications may be true (at least in some respect) (Kolb, 1995). Cognitive reserve is highly discussed in several neurological diseases to explain why some of the patients with large visible brain damage in the magnetic resonance imaging (MRI) scans perform quite normally in many tests, whereas other patients show severe functional losses in independence (although their MRI scans look quite normal). Regeneration of nervous cells is possible also in adult brains and sprouting of dendrites is the standard response of the brain after suffering from an injury.

Still, the ability of the brain to reorganize completely is restricted. A first limitation concerns the growth of nervous cells and dendrites. Nervous cells are produced in the choroids plexus of the ventricular system of the brain. If they should replace neurons that have died due to a stroke, for example, they must migrate across distances ranging from millimeters to centimeters. This migration is unguided for its direction and must surmount the lesion scar. In the case of dendritic and axonal sprouting or growth, the situation can be even worse. If, for example, the connection between the hippocampus and the parietal cortex is lost through a stroke, the distance is several centimeters. In the case of a focal lesion in the internal capsule the distance between the motor cortex and the spinal ventral root neuron may already be one meter. Hence, sprouting and axonal growth can help to compensate cortical lesions. However, even small subcortical lesions, if localized in strategic neuronal pathways, can have effects that cannot be repaired by brain plasticity due to limitations in growth, directedness, and lesion scars. Accordingly, the prospect for recovery after suffering from such a lesion is limited, whereas the prospect for recovery for even larger cortical lesions may be excellent.

Therefore, full recovery is often an unrealistic aim for neurorehabilitation. This opens the question whether it is possible to *compensate* for the impairment. Actually, a certain amount of compensation is involved in the recovery course of many patients where one might have the (false) first impression of a full restitution. Talking to patients and looking at their response times in experimental tasks shows that some amount of tiredness and cognitive slowing is a very common consequence of neurological impairment. This makes sense if recovery involves different and slightly more effortful processing pathways. There is still another reason why the differentiation between restitution and compensation may be somewhat arbitrary. das Nair and Lincoln (2012) and Dirette, Hinojosa, and Carnevale (1999) compared restitutive training and compensatory training for memory-impaired patients or for patients with visual deficits after acquired brain injury. They found that patients in the restitutive training group started to develop internal strategies to deal with the tasks, i.e., they tried to internally compensate for the impairment. Hence, it is very hard to define clear empirical evidence that would distinguish between compensation and restitution.

Nevertheless, in clinical practice it makes sense to distinguish between these two goals. Restitution then means the recovery of a function to a degree that allows a patient to perform (almost) on the same level as before. Compensation means that the patient will rely on different strategies or will need external help to reach a specific degree of independence (at work, at home, etc.).

The strategy to compensate for an impairment will differ according to the additional processes or instruments that must be introduced to enable the target behavior. If restitution is an impossible achievement the best option to consider is the replacement of the impaired function with another one. There are impairments where such a replacement is possible and is a quite usual goal for neurorehabilitation. If, for example, the lesion leads to an impairment of foot dorsiflexion, the patients must learn to use a lateral circular movement to compensate for the impairment. Another neuropsychological example for internal compensation is the adaptation to unilateral homonymous visual field loss, or hemianopia. These patients may have considerable difficulty in reading, crossing the street, or finding objects on a table. Returning to the work of Adhémar Gelb & Kurt Goldstein on the treatment of brain injured soldiers of the first world, Zihl (1988) and Kerkhoff, Munssinger, and Meier (1994) started to treat patients by teaching them to use eye movements toward the visual field loss to compensate for their impairment. This may sound like a trivial problem, but one should be aware that most of these patients do not experience a conscious visual field loss and only notice that there is an issue after getting into some form of trouble. Therefore, they do not experience an inner need to compensate for the visual field loss by using eye movements. Zihl and Kerkhoff could show that anticipatory and automatized eye movements can improve the everyday performance of these patients considerably. We compared a strategy training for eye movements with a visual restoration training (stimulating at the border of the visual field loss) with an occupational training in a randomized control group (Mödden et al., 2012). After 15 sessions of computerized training, patients with compensatory training performed better in a series of search tasks compared to both the restitution training group and the occupational training group. The restitution training group showed no decrease in the visual field loss (for more successful results with restitution training compare; Bergsma & van der Wildt, 2010; Kasten, Wust, Behrens-Baumann, & Sabel, 1998). Our study argues that instead of restitution training, compensation training should be the first-line treatment strategy. However, patients will not always be satisfied with such a choice of treatment. Only a sufficient decrease of the visual field loss will allow them to drive a car again. If successful, only restitution training would help to achieve that goal, but there is no evidence for a large recovery after a stroke (say for >10 degrees in visual angle). Therefore, there is a need to convince the patients that striving for compensation (and not for restitution) is the correct treatment and such a decision should be legitimized.

Internal compensation does not always work, even if there are strategies available that really improve performance. To give one example, which will be discussed later much more extensively, internal strategies to compensate for memory disorders do not (always) work. During the 1980s and 1990s, research considered whether it was possible to teach memory-impaired patients specific strategies to improve their performance. One of the most plausible approaches in the direction of internal compensation was the use of the methods of loci together with the visualization of items to be learned from a list. The method of loci means that patients should imagine a path or a room, and that they should put the items to be learned at specific locations on this path. Afterwards, for retrieval, he or she mentally walks along the path and looks sequentially for the items at their positions. This method is helpful not only for memory disordered patients but also for healthy subjects. The reason for improving memory performance might be related to systematic cueing: The place is used as a cue for remembering the object, and cueing leads to better retention than free recall. Moreover, the method has a high theoretical plausibility when one considers that memory disorders are often restricted to one domain (such as, e.g., verbal learning and memory). Verbal learning and memory is usually impaired after left hemisphere posterior cerebral artery infarcts if it involves the collateral sulcus or the hippocampus and parahippocampus. In this case, visual memory is working properly due to its lateralization to the right mediobasal temporal lobe. Despite this fact, patients with memory disorders used this method only in experimental settings and when being told to do so. They did not rely on such strategies in everyday performance. Most neuropsychologists concluded that internal compensation is not a realistic goal in memory rehabilitation. This consequence will be discussed later. At this point, it can be stated that internal compensation seemingly must rely on "natural" strategies (such as eye movements in visual field defects and lateral movement in lowered dorsiflexion) to have an impact on everyday performance.

Arguing for the importance of a short-hand and "natural" internal compensation strategy to be used by the patients outside the treatment sessions does not mean that all internal compensation strategies invented by the patients themselves are optimal. Reestablishing the functional use of a paretic hand can be delayed or even prevented by the spontaneous compensation to use only the non-paretic hand. This is the background for one of the most important and well-studied interventions in neurological rehabilitation: the constrained-induced movement therapy (CIMT). The main idea of CIMT is that the impairment of the paretic hand is partly a consequence of a learned nonuse, and that patients have to be forced to use it for several hours a day to achieve a better compensation for their impairment. The principles of CIMT have also been used in aphasia therapy and in holistic approaches for the rehabilitation of young patients with traumatic brain injury, often with positive results.

Maladaptation, or suboptimal adaptation (as in the case of learnt nonuse), is also a question for patients with hemianopia and/or neglect. These patients try to compensate for the impairment by extreme head turning behavior. Head turning moves either the object into the intact hemifield or it moves the neglected side down- or upwards (by turning the head in the lateral direction). Although this type of behavior leads to some instant relief of the problem, it is ineffective in the long run. Ultimately, this behavior can lead to neck pain and prevent the use of eye movements for compensation. These patients must be convinced that they should avoid behavior that seems to be helpful and to invest time to learn other behavior that they do not know yet. As already mentioned before, such decisions, which are the next goals of the neurorehabilitative treatment, have to be legitimized.

Let us consider another example of a type of maladaptive compensation that should be suppressed by therapeutic intervention: Memory-impaired patients tend to compensate their impairments by relying spontaneously on verbal rehearsal during encoding of information. Presenting information to them causes them to repeat it to themselves in a pattern of ever-expanding loops. This strategy begins to fail as soon as the limit of the verbal rehearsal buffer is reached. Fig. I.2.1 (Hildebrandt, Clausing, Janssen, & Moedden, 2007) shows the effect of serial clustering (the rehearsal of presented items in verbal working memory) on wordlist learning in a large group of patients

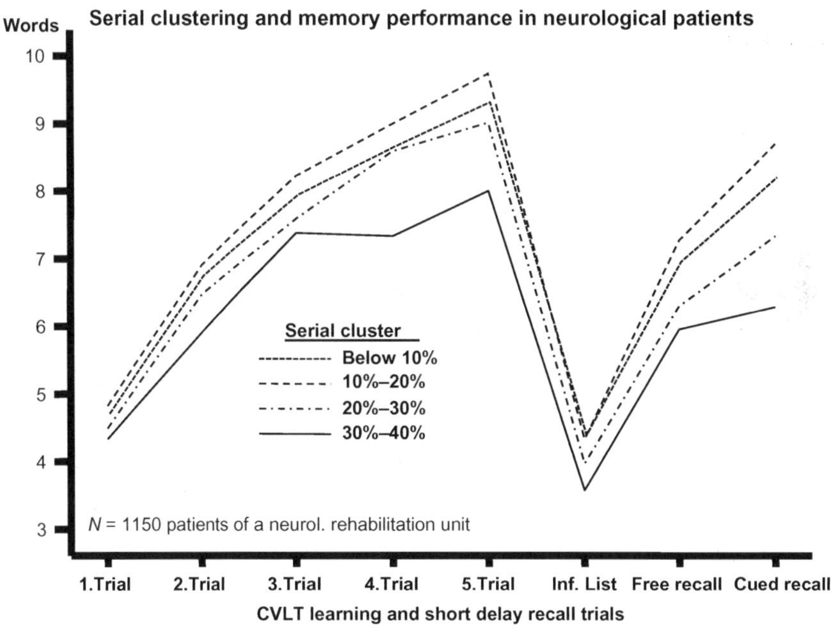

Fig. I.2.1 Effects of serial clustering (rehearsal) on memory performance (Hildebrandt, Clausing, et al., 2007). Patients who try to cluster their recall according to the presentation sequence have lower learning curves.

of neurological rehabilitation (Phase D). The group that tried to cluster (cluster ratio between 30% and 40%) documented a lower learning curve and reduced recall.

Cashdollar, Lavie, and Duzel (2013) showed that this kind of behavior is not only used by memory-impaired patients but also by healthy controls with low memory performance. In their experiment, they presented scenes of places to the participants (healthy controls and patients with bilateral hippocampal lesions). In one condition, the sequence of presentation was interrupted by faces irrelevant for the memory task, while in the other condition there was no interruption. Patients showed better memory performance *with* interruption during the encoding phases. Comparing healthy controls with above-median split memory performance with those with below-median performance revealed a similar result. Cashdollar et al. (2013) also observed increased frontal theta oscillation in the participants benefiting from the interruption during the encoding period. They argue that this theta oscillation may indicate internal rehearsal. Obviously, the participants with low memory performance also used rehearsal as a mnemonic strategy. This usage interfered with deeper encoding and therefore the performance dropped. Presenting the faces interrupted the rehearsal strategy and therefore improved memory performance.

If internal compensation is not possible, then the question of the possibility of an external compensation needs to be clarified. External compensation means the use of aids (e.g., wheelchair instead of walking, prism glasses instead of conjugated eye movements, memory books instead of training of memory consolidation, etc.). The very possibility of external compensation is objectively always good news for neurorehabilitation. This is because in many patients with large acquired brain damage (or with focal but strategic lesions), restitution and internal compensation will not work. On the other hand, on a subjective level, external compensation requires the patient to accept that she or he will be handicapped for the rest of their life. Therefore, offering an external compensation aid implies a statement about the prognosis, which is not always accepted by the patient. In such a case the patient still strives for restitution, while the therapists argue for external compensation. Such a discrepancy between the wishes of the patient and the prospects offered by the therapists leads to the possibility of severe conflicts and a termination of the treatment by the patient. It is important while interacting with the patient that the decision about the next treatment step as well as the decision to finish the treatment can be scientifically legitimized. In this respect, neuropsychological rehabilitation, as part of neurorehabilitation, is part of evidence-based medicine (EBM). EBM is the approach to make health care-related decisions clear and "secure" for patients, health care professionals, and society.

3 DEFINITION OF EBM

EBM is the targeted, precise, and balanced use of current medical knowledge for making decisions about the therapy of individual patients (Sackett, 1997, 1998). The major goal is that the decisions should be made based on individual clinical experiences and the best available external clinical and therapeutic scientific evidence. Such a goal poses a major problem: There is an increasing mismatch between the number of medical publications and the limited time of health care professionals to read the growing number of publications. Therefore, EBM not only strives for the use of more thoroughly scientific knowledge in clinical practices, but also aims for the implementation of pragmatic procedures to transmit a growing knowledge of scientific results to health care professionals. The increasing number of publications about the proper treatment of patients is clearly a problem. Table I.3.1 shows the development of studies registered in PubMed on the topic of neuropsychological treatment. Even the most engaged clinical neuropsychologist will not have been able to follow all of the publications in the two decades after 1990.

Table I.3.1 Development of neuropsychological therapy research: hits of PubMed searches on the term "neuropsychological treatment"

Date range	Number of studies
1960–1970	9
1970–1980	164
1980–1990	1257
1990–2000	4797
2000–2010	13279
2010–2018	28093

EBM therefore has three components: First, to improve the scientific methods used in studies on treatments in the health care system; second, to provide health care professionals with the newest results by a transparent method of summarizing the findings; and third, to help them to apply these results on the level of their clinical knowledge. In the following section we will first focus on the second topic, which concerns the organization of the processes of reviewing and developing guidelines. We will then move on to the first topic, which concerns the methodology of intervention studies. Finally, we will summarize the current knowledge from memory rehabilitation studies. This is done to assist clinical working colleagues in their decisions regarding the treatment of individual patients based on their impairments and the given clinical resources.

4 THE CENTRAL ROLE OF META-ANALYTIC REVIEWS FOR EBM

Commenced in 1993, the Cochrane collaboration is the most well-known international network that produces, updates, and distributes systematic reviews about therapeutic issues (Cochrane reviews). Each review from the Cochrane collaboration follows unified criteria with a fixation of the protocol for the search strategies for identifying interventional studies and for common evaluation criteria of the meaning of these studies. It follows a transparent literature search (in, e.g., Medline, Embase, manually, etc.) and a common and predefined strategy for the evaluation and the meta-analytic presentation of the study results. Elements of the evaluation of the studies concern the methodology of the study, the presentation of the results, the presentation of drop-outs, an evaluation of possible distortion by selection and reporting bias, etc. The reviews are all published in the Cochrane Library (electronic database, CD-ROM, online access). Table I.4.1 shows the meta-analytic summaries for four neuropsychological intervention domains in stroke patients. As one can see from the table, using the very strict criteria of the Cochrane collaboration, neuropsychological interventions in stroke patients are not really based on evidence (but even the Cochrane collaboration admits that the evidence for neuropsychological treatment is better for patients with acquired brain diseases not due to stroke). The specific features of the criteria that the Cochrane collaboration uses to evaluate health care interventional studies will be discussed later. At this point it is only important to note that during the past two decades, the Cochrane collaboration was extremely successful in changing the methodological soundness of interventional studies. To mention only one aspect of their success: To reduce the positive publication bias and the suppression of negative study results, i.e., the tendency of scientific journals only to accept studies with successful intervention, the ClinicalTrials.gov registry was initiated (https://clinicaltrials.gov/). This registry is a worldwide collection of planned, started, and finished public and private clinical studies. Registration is designed to be performed in the stage before they start recruiting patients.

Table I.4.1 Cochrane reviews for functional treatment of neuropsychological disorders in stroke

Study	Meta-analysis
Attention disorders after stroke (July 2013) (Loetscher & Lincoln, 2013)	Many people have problems with attention after stroke. We identified six randomized controlled trials ... The methodological quality of the trials identified and the paucity of studies means that we cannot draw conclusions about the effect of cognitive rehabilitation for attention. More research is needed
Neglect after stroke (July 2013) (Bowen, Hazelton, Pollock, & Lincoln, 2013)	Our review of 23 studies involving 628 participants with stroke found insufficient high-quality evidence to tell us the effect of therapy designed for treating neglect. We did find some limited evidence which suggested that such therapy might be helpful, but the quality of this evidence was poor and more research is needed to confirm this finding
Visual field defects after stroke (July 2017) (Smedslund & Myrhaug, 2017)	At the present time, the evidence for effect of interventions for visual field defects is mostly of very low quality, and the effects are uncertain. An exception involves compensational strategies for which there may be positive effects on visual search, reading and abilities to perform daily activities (ADL)
Memory disorders after stroke (July 2017) (Lockwood, 2017)	Benefits were reported in the short term on subjective measures of memory; however, these did not persist in the long term ... no benefits were reported in objective memory measures, mood, or daily functioning. There was insufficient evidence to support or refute the effectiveness of memory rehabilitation ... This may be because of poor methodological quality of the included studies, inconsistencies in the choice of outcome measures, and small sample sizes. Furthermore, more robust trials of memory rehabilitation that use standardized activity or participatory level outcome measures are required

This helps to analyze whether the positive result of a study is a singular success in a series of failures of the same intervention. Thus, a positive publication bias can be ruled out. Meanwhile, high-ranking journals only accept treatment studies listed at this or similar registration websites, and ethical boards advise applicants to register their trials.

The collection of Cochrane reviews is one of the most important sources for the second step of EBM: the formulation of guidelines for interventions.

5 GUIDELINES FOR TREATING PATIENTS

Meta-analytic reviews like that of the Cochrane collaboration set the background for the development of guidelines, but guidelines are more than just a summary of such reviews. The goal of EBM is the development of a set of scientifically-based rules for the treatment of different diseases and symptoms. Guidelines attempt to synthesize the knowledge from reviews and treatment studies into information regarding how to treat the patients. Usually, guidelines are developed by the national societies of the medical associations. In the case of neurorehabilitation, two examples are the German Society of Neurology and the German Society of Neurology and Rehabilitation. Guidelines are updated every 4 years, focusing on the most recent treatment studies and their results. The goal of guidelines is to evaluate which treatment is the most optimal (through scientifically researched evidence) for the patients. Treatments based on studies with a high methodological standard get a high endorsement ("should be used"), while those based on lower methodological standards get an evaluation of "can be recommended". With even lower evidence, the evaluation of "can be tried" is given, and those which have failed in studies get the evaluation of "should not be used". Interventions that are used without any scientific background (only by clinical or ethical consent), are evaluated as "clinical consensus".

The German guidelines on neuropsychological rehabilitation can be found at the following website: http://www.awmf.org/awmf-online-startseite/awmf-aktuell.html. Cicerone et al. (2000, 2005) reviewed research on neuropsychological rehabilitation with a somewhat different set of criteria as the Cochrane group, but at the same time developed practice standards for the rehabilitation of different neuropsychological syndromes. There is a meta-analytic statistical analysis of the effect size of the studies reviewed by Cicerone et al. According to this analysis by Rohling, Faust, Beverly, and Demakis (2009), the effect size of neuropsychological intervention is 0.3, which is a change in performance of about 0.3 standard deviation toward less impairment.

There is a well-defined hierarchy for the evaluation of treatment studies, which will be introduced in the next paragraphs. Before going into these details, it first should be noted that the guidelines are also evaluated according to well-defined quality standards. These standards are based on three features: the strategy to collect the scientific knowledge, the involvement of all groups involved in the treatment and representatives of the patients, and the transparency in communicating the level to which intervention decisions are based on evidence. High-level guidelines are marked as S3 guidelines, while low-level guidelines are marked as S1. Typically, guidelines include flowcharts concerning what to do with patients suffering from a specific disease or symptom and consider the context for this diagnosis. The guidelines for memory rehabilitation, for instance, propose two different strategies for treatment if the patients are severely impaired, compared to if they are only mildly to moderately impaired.

Guidelines are not laws in a legal sense. To treat patients differently from what is proposed by the guidelines is not forbidden, and even proponents of EBM argue that any treatment decision must be based on clinical knowledge and not only on scientific research. On the other hand, if the chosen interventions do not improve the symptoms and they do not rely on the guidelines, it would be hard for the therapist to assert that everything was done correctly in the event of any conflict with the health insurance provider or the patient.

Irrespective of this juridical question, each neuropsychologist should know the guidelines for neuropsychological rehabilitation, because it is impossible to keep up with the ever-increasing number of publications in this field. As it is impossible to compile a weighted synthesis of these publications as a single individual and because high-standard guidelines integrate the knowledge and experience of all professional groups and the patients, relying on such guidelines is the most reasonable solution to this problem.

6 HOW CAN WE MEASURE TREATMENT EFFECTS?

Up to now, we have focused on the EBM process that starts after treatment studies have been conducted. This process relies on two steps: systematic and meta-analytic reviews and the definition of guidelines. Now we will focus on the question which features of a treatment study qualify it as being high level. Let us start with examining how to measure treatment effects (or outcome of interventions). To ease into the question, an early study on the treatment of traumatic brain injured patients by sensoric stimulation will be presented (Hildebrandt, Zieger, Engel, & Kraft, 1999). Forty-seven patients with at least 3 days of unconsciousness and a Glasgow coma scale (GCS) score of below 8 at admission were recruited for this study. The Glasgow coma scale score was assessed at admission to the intensive care unit and reassessed before the beginning of the treatment. The patients were divided in three groups according to the severity of the coma at admission. Patients with a GCS score of 3–5 belong to the severe coma group, patients with a score of 6–7 to the moderate coma group, and patients with a score of 8–10 to the weak coma group. The patients were systematically presented with a series of visual, tactile, and acoustic stimuli for periods of 5 min. We measured the heart rate variability to see whether the stimulation would lead to an increase of the sympathetic tone and the coupling between the heart rate and the muscle tonus at the forearm. This was done to analyze the integration of the motor system and the autonomous nerve system (cardiosomatic coupling) as a response to our stimulation. Our patient group was severely impaired and the GCS at admission was on average 4.3.

The results showed that cardiosomatic coupling increased with higher GCS scores and that in the group with the highest GCS score, the coupling increased in response to the multimodal stimulation (Fig. I.6.1). The analysis of the heart rate variability yielded a similar result: The sympathovagal ratio increased in correspondence to our stimulation for the group with a moderate GCS score and especially for the group with a high GCS score. We concluded that multimodal stimulation improved cardiosomatic coupling and the sympathovagal balance, and that it therefore might improve outcomes in patients with severe brain injury.

The study had several obvious methodological issues. To name only one, we did not measure the long-term outcome after several months. On the other hand, we were clearly aware that one multimodal stimulation will not change the clinical outcome in such a group of patients. However, there are also several other questions concerning the measurement of the outcome of an intervention that can be discussed using this study. The first question is: What kind of improvement in outcome qualifies as proof that this intervention should be integrated into the set of recommended and socially funded treatments? In the study, we focused on two purely experimental parameters: increase of cardiosomatic coupling and increase of the sympathovagal balance. Both were derived from physiological knowledge, but both are also not direct measures of the primary goal of neurorehabilitation in this patient group: regaining consciousness and voluntary interaction with the environment. Thus, for the evaluation of interventions, one should distinguish between experimental outcome parameters and clinical outcome parameters. Experimental outcome parameters do not qualify as parameters that argue for an integration of a successful intervention in routine treatments. To show that such interventions are meaningful they must also change the *clinical outcome*. There is only one exception to this rule: In the case that the correlation of the experimental parameter (e.g., blood pressure) and the clinical parameter (cardiac disease) is very high, the experimental parameter

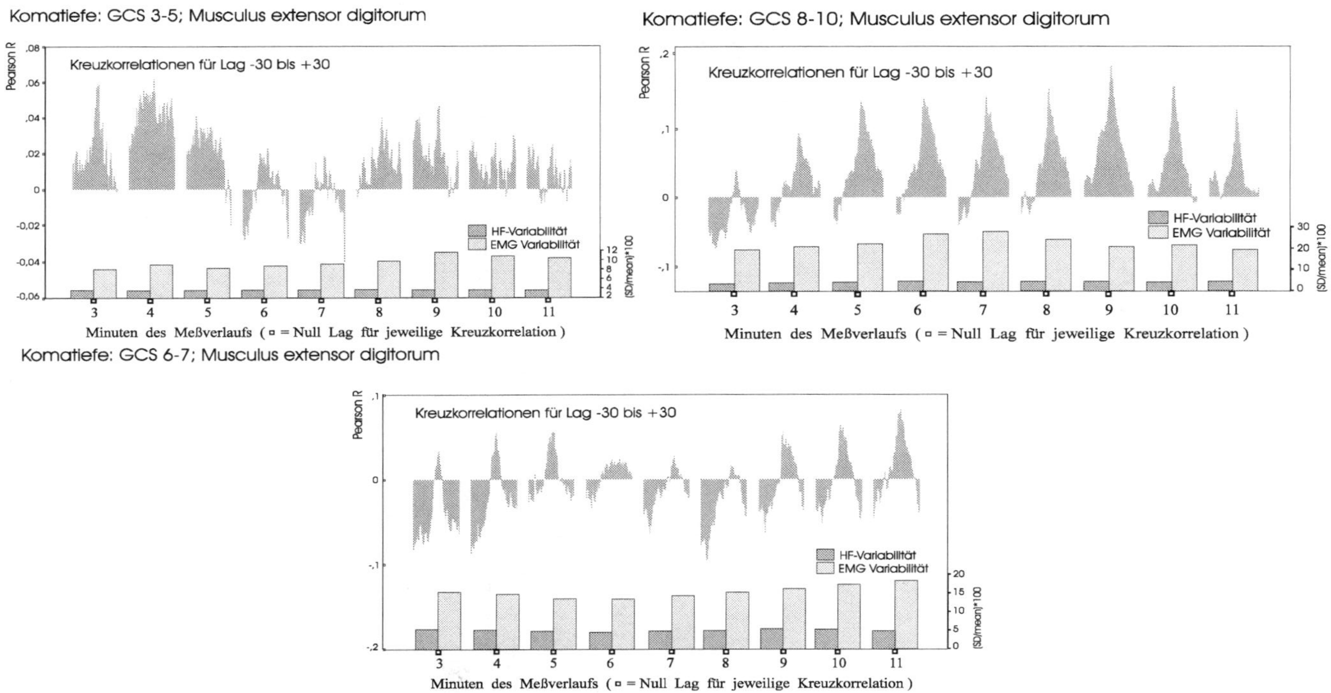

Fig. I.6.1 Cardiosomatic coupling in TBI patients with different depths of coma (Hildebrandt et al., 1999). The figures show that cardiosomatic coupling decreases with coma depth and at the same time increases due to coma stimulation, but this is only true in patients with moderate or mild coma depth.

might be accepted as a surrogate marker. Nevertheless, for neuropsychological rehabilitation there are no surrogate markers established. This directly leads to the next question: What is a clinically relevant change in outcome? There are many answers to this question, depending on the disease and its consequences. For example, in the treatment of cancer a good clinical outcome measure seems to be the mortality rate. In stroke treatment, the duration of time to the next stroke is another accepted outcome parameter. In the treatment of back pain, one might look at days of absence from work. In neurorehabilitation, outcome measures focus on the reduction of disability, i.e. the ratio of patients that must live in a nursing home, the need for care if living at home, the rate of re-employed patients, etc. There are also international scales that are accepted as outcome measures. Examples are the FIM (functional independence measure) or the Barthel Index (BI), which tries to quantify the ability of the patient in different domains (FIM) or the time that is spent in caring for the patients in these domains (German version of the BI).

There are three other aspects of outcome measures that should be mentioned. First, the cost-benefit analyses for treatments have recently gained an increased importance. For example, in the evaluation of drugs, a higher price for a new drug is only accepted if the drug performs better than an already existing one. There are also some qualitative dimensions in evaluating the evidence of a specific intervention. The first dimension is the global clinical impression score. This score functions as an outcome variable in many studies that use drugs in the treatment of mental health disorders and is assessed by a physician who is unfamiliar which of the patients receive the drug. On the other hand, *quality of life* of the patients is not an accepted outcome measure to show the effectiveness of an intervention. Excluding quality of life from the outcome parameters may be criticized for several reasons and it is often argued that quality of life should be considered the central aspect of an intervention. If this would be correct, quality of life should be included in any evaluation of interventions. On the other hand, quality of life is a notoriously subjective experience and it should be noted that it may be higher for patients who are behaving in a manner that actually increases their risk for suffering from diseases (for instance by smoking). In rehabilitation after brain injury, Kurt Goldstein was the first to notice that patients seem to look for an ecological niche where they can live optimally with their impairment. Unfortunately, the representation of the impairment in the mind of patients (and subsequently, the possibility for improvement) is not always aligned with reality. This misrepresentation may be associated with considerable burden for patient's relatives. With that in mind the treatment is directed against such suboptimal adaptation to the disease consequences, lowering the experienced quality of life (at least in the short run). Therefore, improvement in quality of life is not a clinical outcome parameter that is accepted as justification for adding an intervention to clinical practice. Nevertheless, it is clearly an important aspect in neurorehabilitative treatment studies, which always depends on the motivation of the patients.

As Table I.6.1 shows, evidence of treatment success is usually weighed on at least three levels. Such evidence may concern *experimental* variables such as muscle strength, response time, or changes in blood-oxygen-level dependent (BOLD) response when conducting a pre-/postfunctional MRI scan. These variables may also be grouped under the label of *impairment*. If they concern clinical variables or clinical assessments, for example measured with scales (FIM, BI), global clinical impression of health care professionals, time to commence working again, or the view of caregivers on independence in (instrumental) activities of daily living, they are called *disabilities*. Evidence of treatment can also be measured in *health economic terms* such as cost-effectiveness and cost–benefit analysis (comparing two similar interventions). The meaning of these outcome parameters differs depending on the social position: Someone who works for a health insurance company will be interested in cost-effectiveness studies. For a health insurance company, the fact that 187 people must be treated with acetylsalicylic acid (ASA) to be sure that one lethal heart attack is prevented may make sense when comparing the costs of the drug with the cost of treating a heart attack patient. It is not so clear if a patient would want to take ASA knowing that the chance that the drug would be helpful (or its number needed to treat) is 1:187. Patients are primarily interested in the evidence that a specific

Table I.6.1 Different qualitative levels to demonstrate treatment effects

When to talk about evidence-based treatment
Significant effect of intervention shown in: • Experimental variables e.g. fMRI or reaction times (often termed ***impairment***) • Clinical variables concerning functional **disability** (mortality, changes in *disability* scales such as the FIM, the BI, global clinical impression, etc.). • Health economic variables e.g. cost-effectiveness and cost–benefit analysis oriented on "social handicap" (duration and cost of treatments, job loss, caregiving costs etc.).

intervention will cure their disease and increase their quality of life. This is similar to the view of health professionals, who often act on their tacit clinical knowledge. Still, as mentioned previously, the goals of health professionals and patients may differ, depending on how patients have adapted to their disease. Last, but not least, experimental research on disease and rehabilitation may be quite uninteresting in respect to the short-term clinical consequences, but very interesting regarding the long-term development of better interventions.

7 HOW TO RATE NEUROPSYCHOLOGICAL ASSESSMENTS AS OUTCOME MEASURES

Do neuropsychological test results correspond to an experimental finding such as a change in muscle tonus or a difference in BOLD signaling? Or do they correspond to a clinical outcome parameter? In other terms: If walking is the outcome parameter for gait training and walking is measured in the time it takes to reach 10 m, is the performance in a neuropsychological memory test comparable to such a measure? There are no easy answers to these questions. On the one hand, many colleagues have argued that an improved performance in memory tests is irrelevant in clinical terms, as it provides no proof that the patients will perform better in their everyday lives. This view argues that neuropsychological tests are experimental outcome measures but not clinical ones (for two studies that demonstrate how restitutive memory training – even when it is ineffective in terms of improving test results – may nevertheless lead to the same increase in reliance on external memory aids as seen in more behavioral compensatory interventions, compare das Nair and Lincoln (2012) and Dirette et al. (1999)). On the other hand, only a few scientists ask whether an increased speed in walking an experimental distance after gait training really translates into everyday performance at home or work. To give one example: Corbetta, Sirtori, Castellini, Moja, and Gatti (2015) performed a Cochrane review on the effectiveness of CIMT, one of the best evaluated interventions for motor rehabilitation. They found that most of the study focused on the impairment level, determined by Wolf motor function test, Fugl-Meyer assessment, hand strength, etc. Patients profited from CIMT on this functional level, although the effect size was low (about 0.34). About 25% of the studies also involved the disability level, and all used either the FIM or the BI. The review showed that the effect of the treatment on this disability level was not significant. Interestingly, in their summary, Corbetta et al. (2015) do not really question the effectiveness of

CIMT or insist on more trials but rather focus on the impact of patients' characteristics and outcome:

> *CIMT is a multi-faceted intervention where restriction of the less affected limb is accompanied by increased exercise tailored to the person's capacity. We found that CIMT was associated with limited improvements in motor impairment and motor function, but that these benefits did not convincingly reduce disability ... Information about the long-term effects of CIMT is scarce. Further trials studying the relationship between participant characteristics and improved outcomes are required.*

Now compare that statement to that of Piras, Borella, Incoccia, and Carlesimo (2011) (or even more pronounced, das Nair, Cogger, Worthington, and Lincoln (2017), Table I.4.1) reviewing the effect of memory training:

> *When only randomized controlled trials (RCTs) are considered, the intervention improved the trained skill in all the reviewed studies, while a 44% demonstrated a positive effect on performance in memory tasks other than the trained one, and only 20% of studies reported some improvements at the disability level.*

Therefore, in the physical domain of rehabilitation, experimental parameters that are quite similar to neuropsychological test results are usually accepted as relevant outcome parameters. In this domain the insistence to demonstrate significant effects on the disability level is much lower than in neuropsychological rehabilitation.

Nevertheless, showing that patients improve after a treatment in a word list learning test and then arguing that the treatment was effective is intuitively not enough to show the clinical evidence for that treatment. Neuropsychologists involved in memory rehabilitation have responded in different ways to deal with that objection. One strategy is to show that neuropsychological tests have ecological validity. This can be done by adapting the tests to include tasks that recreate situations where memory is necessary in activities of daily living. Using text reproduction instead of word list learning is one obvious and easy method to increase the ecological validity of memory testing, because text reproduction is something people do very often in daily life (for example reading a newspaper). To develop a test that includes several subtests which represent situations in everyday life would be another step in the direction of ecological validity. One such test that attempts this is the Rivermead Behavioral Memory Test. Ecological validity can also be demonstrated by predicting the psychosocial status of the patients using their memory tests results and controlling for other important variables. There are several studies that demonstrate that low memory performance in the California Verbal Learning Test predicts job loss after acquired brain damage and dependence on care, i.e. not being able to live alone

(Chernoff, Martin, Schrock, & Huy, 2010; Dikmen, Machamer, Powell, & Temkin, 2003; Dusankova, Kalincik, Havrdova, & Benedict, 2012; Hanks, Jackson, & Crisanti, 2016; Morrow et al., 2010). In this respect, one may consider the results of these tests as a surrogate marker of everyday performance (as, for example, the number of new lesions visible on a MRI scan counts as a surrogate marker for disease progression in multiple sclerosis). It would be very helpful for neuropsychological rehabilitation research if more studies on the relevance of such tests for the prediction of job status and care dependence were available. If this were the case, then clinically relevant changes and cut-off scores could be calculated that would allow the division of patients into groups with and without such improvement.

Yet another strategy is to use rating scales and to ask relatives or colleagues if memory performance has increased due to the treatment. Although this may look like the "gold standard" for showing clinically relevant improvements, one should keep in mind that it is not so easy for lay people to distinguish between attention, memory, and unconscious adaptation to living conditions. Therefore, such rating scales might suggest or miss relevant changes, just because solving cognitive tasks often relies on many different subroutines. Another problem with using rating scales is their "ecological validity" for measuring memory impairments. The FIM and the Barthel Index, often used in neurorehabilitative research, are focused on activities such as mobility, self-care, eating etc. Memory does not play a major role in being successful in these types of tasks. Questionnaires that focus on very specific problems resulting from memory impairment may indicate strong improvements, but only in very few aspects (see the next paragraph). Rating scales for instrumental Activities of Daily Living (iADL scales) such as that of Lawton and Brody (1969) seem to cover tasks where memory plays an essential role (using a telephone, shopping, preparing a meal, managing money, traveling). The advantage of such iADL scales is that they are broadly accepted in the scientific community. Therefore, should an intervention show an improvement in iADL scores, it will convince many colleagues of the meaningfulness of that intervention. However, for younger patients with only mild memory impairments, these kinds of tasks will pose no major problem and the focus should be on work performance or success in learning at school or university.

Fig. I.7.1 shows two additional strategies to deal with the problem of clinical relevance: One is more related to theoretical assumptions from cognitive neuropsychology, while the other is related to cognitive behavioral therapy (CBT) (Wilson, 2013). The first strategy reframes the problem of validity and relevance in three substeps. The first step concerns the

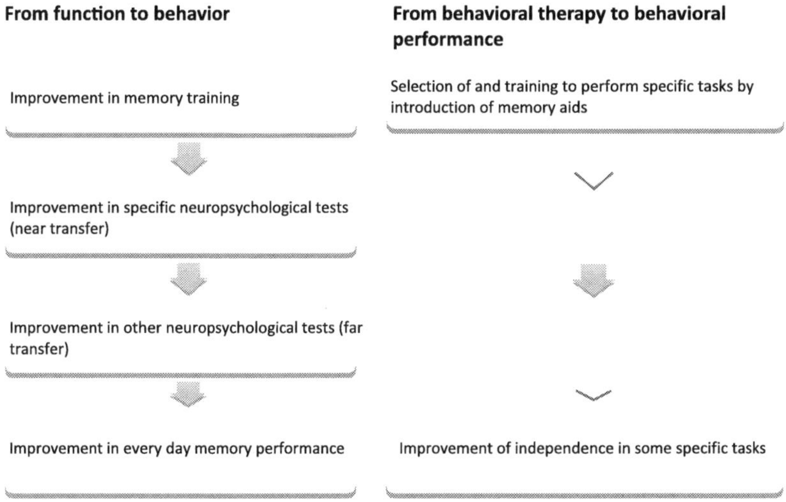

Fig. I.7.1 Suggested steps of successful neuropsychological memory rehabilitation.

generalization of an improved performance in a trained task to neuropsychological tests, assessing the trained ability. The second step would be an additional improvement in cognitive domains not trained. Finally, the third step involves a generalization to everyday performance, assessed, for instance, by questionnaires on the impaired functions and rated by significant others. The CBT approach ignores the goal of improving memory as a cognitive function and instead argues for focusing on goals that the memory-impaired patients would like to achieve. The task of neuropsychological rehabilitation then is to help the patients achieve these goals, often by learning strategies that avoid any load on episodic memory (e.g., errorless learning as a shaping method). In this case the ecological validity seems to be part of the treatment, not a problem that has to be solved by additional tricks or measurements. Although this might at first appear to be a clever solution to the problem of ecological validity, a second look shows that this is only correct when assuming that the environment is very static, the goals of the patients always remain the same and such goals are realized in very similar situations. In other words, teaching a patient to move from A to B may be very helpful if the patient is often (always) in A, but it is insufficient if the patient is often in C or D, etc. Just focusing on one or two goals of the patient is justified if the patient is severely memory impaired, but this approach is confronted with the ecological problem of context-specific goals depending on each specific situation (a problem that can only be solved with a residual function of episodic memory).

Some authors have used *self-rated* questionnaires as primary outcome parameters for neuropsychological rehabilitation research (das Nair & Lincoln, 2012). The background for this decision seems to be the view that such questionnaires (if focused on everyday memory performance), combine the feature of being "ecologically" valid (therefore covering far transfer effects) with the perspective of the patients receiving the treatment. There are several problems with this decision, as the use of self-rated questionnaires is not an alternative to show objectively-measured changes. The first problem is that patients rating these questionnaires know that they are being treated. This will have an impact on the results even if the patients are uninformed of their treatment status. Berg, Koning-haanstra, and Deelman (1991) published an influential paper on memory rehabilitation. There were three different groups; one with a compensatory training, one with drill and practice training, and one untreated control group. When patients were asked after the training if it was helpful and if it could be useful in everyday situations, the two active treatment groups answered with yes, whereas the non-treated group answered with no. Neuropsychological testing revealed that as the study predicted, only the group with compensatory training showed improvement in their memory performance. In the drill-and-practice group there was no such change. As just giving patients tasks that lead to the *impression* that memory rehabilitation is taking place changes their view on their performance, self-rated questionnaires are not suitable as (primary) outcome parameters.

8 LEVEL OF EVIDENCE

In the previous section, it has been argued that only improvements in clinically meaningful parameters qualify as successful outcome measures, and that such measures will lead to the acceptance of a specific treatment as an evidence-based intervention to improve health. The next question to address is how we can design interventional studies so that they qualify as high-level studies. There is a clear hierarchy of evidence levels in the evaluation of studies that can be criticized for different reasons, but which still are the accepted standard in their fields and cannot be ignored. This hierarchy starts at the bottom with the consensual agreement of experts in the field (Table I.8.1). Having a common consensus without having any actual studies that support it falls under the lowest level of evidence (Level IV). Please note that many interventions in neurorehabilitation have not been a topic of randomized controlled studies and yet these interventions are provided routinely in

Table I.8.1 Levels of evidence-based medicine

Level	Study characteristic
Ia	Meta-analysis of randomized controlled trials (RCTs)
Ib	At least one RCT
IIa	At least one controlled and not RCT
IIb	At least one different type of an experimental trial (unspecific control group, SCED, this latter design is sometimes also rated as level III)
III	Good descriptive not-experimental trial (correlative study, single-case pre-, post-study)
IV	Expert report, authoritative opinion

clinical practice (and nobody has any problems with them). To mention just two examples: As far as I know, no studies have been conducted to see if the long-term effects of offering wheelchairs to patients unable to walk after a unilateral stoke are superior to not providing wheelchairs. There are also no RCTs to determine the most effective type(s) of antidepressants for patients with post-stroke depression and large lesions. Nevertheless, in both instances, one will not hesitate to offer the patients the external help (such as the wheelchair) or the drug. The next level (Level III) concerns single-case observational studies: In this situation, something was done with the patients and the professionals noted a positive response, but this was only on an observational basis (the intervention was not withdrawn to see whether the positive response declined). Conducting interventions with an experimental design yields the next level of evidence (Level IIb). Interventions analyzed in a single-case experimental design (SCED; discussed below), belong to this level, but also studies using control groups that are not matched or not selected for the specific intervention study. Level IIa concerns controlled studies where the patients belonging to the group that did not receive treatment were not selected randomly. For example, if the intervention was done first and the control group was examined afterwards, this would be a Level IIa study. Another possibility is that the two patient groups were matched for specific features (e.g., gender, age, and education) but not for other features that may have played a role. Matched controlled studies also do not reach the highest level of evidence.

The highest level of evidence is yielded by RCTs (Level Ib). Such studies have several additional features, which will be discussed extensively below. Randomization is the only systematic strategy to avoid any kind of bias in the groups, if it is unknown which factors may bias the results in a specific direction. Randomization works properly only in studies with a large number of patients. Therefore, the core evidence of EBM is based on large patient trials. One can also randomize patients into small groups, but this always implies a high risk that a selection bias will be present between the groups. As a result, this may imply that the positive study results are worthless because one does not know whether they stem from the selection bias or from the intervention. Power calculation is another factor that increases the number of recruited patients and we will come back to this later in the chapter.

A properly designed and well-executed RCT reaches a level of evidence of Level Ib. However, there still remains the possibility that the result of one trial might be a chance finding. Moreover, almost all studies have some methodological problems, even if they qualify as RCTs. Therefore, only a series of RCTs with positive results provides the firm knowledge that this intervention is evidence-based. Consequently, the highest level of evidence is reserved for a meta-analytic review of different RCTs in which most of them yielded a positive result.

9 THE RCT AS "GOLD-STANDARD" IN REHABILITATION RESEARCH

There are three major and several additional features that qualify whether a study can be considered a high-level RCT. The three major features concern the strategy to guarantee the *structural equality* of the involved patient group, the strategy to analyze the data (*evaluation equality*) and the blinding (*observational equality*). The additional features concern the transparency in the recruitment of patients and the reasons why some people withdraw from the study if all data are reported completely.

9.1 Structural Equality in RCTs

Structural equality can be split into developing randomized groups and in avoiding any bias in allocating the patients to these groups. Randomization means that patients will be allocated in one of two or more groups irrespective of any other features of the patient and only through the randomization protocol. This is not always perfect, as there are several strategies that might be seen as randomization but which in practice will not avoid a structural bias. Pseudo-randomization (for example in accordance with name, day of admission) does not work properly. The main problem of pseudo-randomization is that it follows a rule and this rule is foreseeable. Therefore, the patient, or the one who recruits the patient, is able to influence whether he or she will be a member of the treatment group or the control group. Therefore, randomization must be done before starting the study and should employ a strategy that is not intelligible for the people involved in the study. The question then turns to how we can accomplish this randomization. One easy possibility is to generate randomized groups using software such asSPSS or a program such as Excel. As a low-tech solution, coin tossing also works.

Constructing a list that randomly allocates patients to either one of two or more groups is only half of the work to guarantee structural equality. The other half is to guarantee the concealment of this list. In other words: The person who recruits the patient has be unaware of the assignment of the patient to the group (verum, placebo). Only after the recruitment can the assignment of the patients according to the randomization list be done by phone, a series of covert letters with assignment, etc. Why is concealment so important? Let us have a look at an old study conducted to show that patients with increased bleeding after suffering a physical trauma have a higher chance of survival if aprotinin (a drug which may reduce bleeding) is added to routine administration (Table I.9.1). A first look at the table might give the impression that the study was highly successful. Compared to treatment as usual (8.4% plus 11.5%), patients treated additionally with aprotinin (11.6% plus 21.3%) had a higher survival rate. However, a second look at the table shows that the study had several methodological problems. The main problem is that a fairly high number of patients were not treated should have been treated (222 + 588, i.e. about 1/6 of all patients). This is not the only problem. The supposed difference in efficacy between the two treatments is based on wrong assignment in the RCT (compare the left upper and the right lower cell of Table I.9.1). Patients that were treated according to proper randomization did not differ in their frequency of suffering a bleeding (left lower and right upper cell of the table). What may have

Table I.9.1 A trial with aprotinin to reduce bleeding events

N (14 days survival rate, percentage)	Even day numbers at admission	Uneven day numbers at admission
Treatment as usual	222 (8.4%)	1909 (11.5%)
Treatment as usual and aprotinin	1964 (11.6%)	588 (21.3)

Note that several patients were not treated as they should have been treated, and the considerable difference between these sets of numbers.

happened? The physicians recruiting the patients were convinced that aprotinin is an effective drug. Therefore, they "randomized" (for example) younger patients and those with a suspected better clinical outcome more often to the aprotinin treatment group (right column and second line of the table). The study shows that physicians were quite able to predict which patients will survive, but the study does not show that aprotinin is effective in reducing the mortality rate.

The problem with pseudo-randomization or any foreseeable randomization key (= randomization without concealment) is that there may be different selection mechanisms between the groups that are influenced by the clinical knowledge of the people recruiting the patients. If these people are convinced of the new treatment and have a tacit knowledge about who should take the new treatment and who should not, incorrect assignment will occur. As such a bias cannot be verified for each patient, any correction of this bias is impossible. A lack of concealment has a large impact on the results of a study, which can produce an even larger impact than a lack of blinding. Schulz, Chalmers, Hayes, and Altman (1995) analyzed studies with adequate concealment, with unclear concealment, and without concealment. The effect size of unclear concealment was about 1/3 and went up to 40% in studies without concealment, compared to studies with adequate concealment. Therefore, concealment is a very important aspect of RCTs and (together with randomization), guarantees that both groups exhibit structural equality.

9.2 Evaluation Equality

Concealment helps to avoid the risk of wrongly assigning patients to treatment arms to which they are not randomly assigned. Still, there might be patients who get assigned to the "wrong" treatment, i.e. verum treatment for patients assigned to placebo, and vice versa. This leads to the question of how to statistically evaluate studies with wrongly assigned patients. This is the core question for evaluation equality. There are two accepted strategies to deal with wrongly assigned patients. The first strategy employs the *intention to treat* (ITT) analysis. This is a very tough strategy because it evaluates the effect of the treatment according to the treatment that the patient should have received. This means that patients that should have been verum patients but instead received placebo treatment will count as verum patients. This is despite the fact that they did not receive a treatment (and vice versa). The reason for this strategy is that there might have been a selection bias in group assignment and this selection bias might explain the significant impact of the treatment. To correct the selection bias, the originally intended (and "correct") patient assignment is used for the statistical evaluation.

On the other hand, such an evaluation strategy will reduce the impact of the treatment. This is because patients that never received treatment will count as treated patients. The other way to deal with wrongly assigned patients is the evaluation *per protocol*. In this case, all patients that have not been treated correctly will be excluded from the statistical analysis. Please note that this strategy is not completely free from bias: If the incorrect assignment was based on a clinical bias (not including patients into the intervention group because they do not look like a patient that will profit from the intervention) then the evaluation per protocol will not be fully effective in correcting the bias. Table I.9.2 shows the result of a study on bypass operations in the treatment of coronary heart disease. The question was whether bypass operations would improve survival time. Again, several patients were not assigned to the correct group (50 + 26). The new treatment (bypass operation) group had a much lower mortality rate than the treatment-as-usual group (4.1% compared to 9.5%). This advantage was lower when evaluating the study per protocol and even lower when evaluating the study with the ITT strategy (5.3% compared to 7.8%).

The figures of Table I.9.2 show again that more patients were treated with the new method and that these patients had a higher survival rate than the patients treated wrongly in the "as-usual" treatment method. Knowing what might help and knowing what to do to get patients in this group will always lead to distortions in the allocation of patients. It follows that, to avoid an unknown bias, all patients included in the study should be evaluated

Table I.9.2 Another trial with wrongly assigned patients and the consequence for the treatment effect depending on the strategy of evaluation

Planned assignment	**As usual**		**Verum treatment**	
Real treatment	As usual	New treatment	New treatment	As usual
N	323	50	368	26
Died	27	2	15	6
As treated	9.5% (33/349)		4.1% (17/418)	
Per protocol	8.4% (27/323)		4.1% (15/368)	
Intention to treat	7.8% (29/373)		5.3% (21/394)	

according to the randomization code (ITT). This should be the primary evaluation strategy and it ensures evaluation equality. The secondary evaluation strategy is "per-protocol (PP)", which requires a restriction in the analysis to only the patients that conformed to the protocol (exclusion of all patients violating the protocol). If there are greater differences between ITT and PP then they must be discussed, because of a possible selection bias.

9.3 Observational Equality or the Problem of Blinding

A double-blind study means that neither the patients nor the people treating and assessing the patients are informed about the group status of the patients (verum vs placebo). By comparing studies with and without adequate blinding, Schulz et al. (1995) noticed a difference in effect size of 0.17. Placebo treatments do lead to changes in outcome and these changes differ from binary and continuous variables (Table I.9.3). The impact of placebo also depends on the type of symptoms being treated. In any case, the effectiveness of placebo argues that any specific treatment effect has to be compared with the effect of the global effect of placebo. This is a real problem when non-pharmacological interventions as provided as rehabilitation efforts, because it is hard for such approaches to design a placebo treatment.

In the meta-analytic review of Rohling et al. (2009) (Box I.9.I) the effect of placebo or spontaneous recovery in neuropsychological rehabilitation was even higher, i.e. up to an effect size difference of 0.4. On the other hand, Rohling et al. (2009) also included SCED studies and in these studies the placebo effect might be a little bit higher (due to several reasons).

Double-blinding is enough in drug studies, but not in neuropsychological rehabilitation studies. Because placebo can improve the performance of the patients (for instance just by increasing their motivation), the comparison between a waiting control group and the verum group is not enough to show an advantage of the neuropsychological treatment. The significant treatment effect might be due to the attention given to the verum group and the increased motivation of this group. Therefore, double blinding in neuropsychological rehabilitation research means that the placebo group is also involved in some common activities. To incorporate a placebo group that is just doing *something* will also not work, because the patients will notice that the group they are a member of is not the treatment group. The same holds for the therapist who is delivering the placebo treatment. For neuropsychological rehabilitation studies, we even have to consider

Table I.9.3 The impact of placebo on binary and continuous outcome measures (Hrobjartsson & Gotzsche, 2001)

Effect of placebo in trials with binary or continuous outcomes				Effect of placebo on specific clinical problems			
Outcome	**No. of participants**	**No. of trials**	**Pooled relative risk (95% CI)[a]**	**Outcome**	**No. of participants**	**No. of trials**	**Pooled relative risk (95% CI)[a]**
Binary				Binary			
Overall	3795	32	0.95 (0.88 to 1.02)	Nausea	182	3	0.94 (0.77 to 1.16)
Subjective	1928	23	0.95 (0.86 to 1.05)	Smoking	887	6	0.88 (0.71 to 1.09)
Objective	1867	9	0.91 (0.80 to 1.04)	Depression	152	3	1.03 (0.78 to 1.34)
			Pooled standardized Mean difference (95% CI)[b]				Pooled standardized Mean difference (95% CI)[b]
Continuous				Continuous			
Overall	4730	82	−0.28 (−0.38 to −0.19)	Pain	1602	27	−0.27 (−0.40 to −0.15)
Subjective	3081	53	−0.36 (−0.47 to −0.25)	Obesity	128	5	−0.40 (−0.92 to 0.12)
Objective	1649	29	−0.12 (−0.27 to 0.03)	Asthma	81	3	−0.34 (−0.83 to 0.14)

[a]The relative risk was defined as the ratio of the number of patients with an unwanted outcome to the total number of patients in the placebo group, divided by the same ratio in the untreated group. A value below 1.0 indicates a beneficial effect of placebo.
[b]The standardized mean difference was defined as the difference between the mean values for unwanted outcomes in the placebo and untreated groups divided by the pooled standard deviation. A negative value indicates a beneficial effect of placebo.

BOX I.9.1 Abstract of Rohling et al. (2009).
Effectiveness of Cognitive Rehabilitation Following Acquired Brain Injury: A Meta-Analytic Re-Examination of Cicerone et al.'s (2000, 2005) Systematic Reviews

The present study provides a meta-analysis of cognitive rehabilitation literature ($K=115$, $N=2014$) that was originally reviewed by Cicerone et al. (2000, 2005) for the purpose of providing evidence-based practice guidelines for persons with acquired brain injury. The analysis yielded a small treatment effect size (ES = 0.30, d'' statistic) directly attributable to cognitive rehabilitation. A larger treatment effect (ES = 0.71) was found for single-group pretest to posttest outcomes; however, modest improvement was observed for nontreatment control groups as well (ES = 0.41). Correction for this effect, which was not attributable to cognitive treatments, resulted in the small, but significant, overall estimate. Treatment effects were moderated by cognitive domain treated, time postinjury, type of brain injury, and age. The meta-analysis revealed sufficient evidence for the effectiveness of attention training after traumatic brain injury and of language and visuospatial training for aphasia and neglect syndromes after stroke. Results provide important quantitative documentation of effective treatments, complementing recent systematic reviews. Findings also highlight gaps in the scientific evidence supporting cognitive rehabilitation, thereby indicating future research directions.

triple blinding: the blinding of the patients, the blinding of the assessor, and the blinding of the therapist. A therapist who knows that all the work he is doing counts as placebo will clearly show different motivation than the colleague who knows that the results of the study depend on her or his work.

There are only two solutions for this specific problem of blinding in neuropsychological rehabilitation research. The first solution is to use active control groups as the placebo group, i.e. interventional groups that receive either a treatment in another cognitive domain or the standard treatment, which can then be compared to the verum treatment. Please note that showing a significant difference between "treatment as usual" and the new treatment is much harder to accomplish than between the new treatment and a placebo treatment. Most of the studies on new anti-depressants would have failed if the drugs were compared to the effects of

BOX I.9.2 Cochrane Review on Fluoxetine (Magni et al., 2013).
Cochrane Review on Fluoxetine
In May 2012 we searched, in a wide-ranging way, for all the useful studies (randomized controlled trials, or RCTs) we could find which compared fluoxetine with any other antidepressant in treating people with depression. One hundred and seventy-one RCTs were included, with 24,868 people in the analyses. Combining the results from all the trials, fluoxetine was similarly effective, but better tolerated, than older generation (tricyclic) antidepressants. In comparison with other new generation antidepressants, important differences in efficacy and in tolerability were found between fluoxetine and some of the antidepressants, for example, fluoxetine was less effective than sertraline and mirtazapine but better tolerated than reboxetine. These differences might have a clinical impact in everyday practice. However, when interpreting these differences, it is important to bear in mind that the studies were short in duration (eight weeks or less) and that the average size of each trial was small (each included around 100 people).

older drugs and not to placebos (see the Box I.9.2 on fluoxetine). Nevertheless, making a comparison to the standard treatment is the only strategy for solving the blinding problem effectively. One inferior solution to this problem is to compare the verum treatment with a treatment of a different cognitive deficit. In this case the patients might ask themselves why the assessment is related to functions not targeted by the treatment they believe they are getting. Another option is to compare the verum treatment to the same but less intense verum treatment. The latter solution is often used in rehabilitation research. However, it has the disadvantage that there is no a priori knowledge as to how many treatment sessions must be delivered to see the first increases in performance or how to motivate the patients during the outcome assessment when they have received only some treatment compared to many treatment sessions.

Schulz et al. (1995) analyzed which of the three aspects of equality (structural, evaluation, or observational) has the main impact on the effect size documented in RCTs. The strategy of randomization and of excluding patients did not remain in the regression analysis and the main effect relied on problems in adequate concealment and blinding (Table I.9.4). This should be remembered when planning or reading a RCT.

Table I.9.4 Different aspects of RCTs and their impact on treatment efficacy if violated (Schulz et al., 1995): Association Between Four Dimensions of Methodological Quality and Estimates of Treatment Effects in the 229 Adequately and Unclearly Concealed Trials[a]

Measure of methodological quality	Ratio of odds ratios (95% confidence interval)	X^2 (df)	*P*
Allocation concealment	–	–	–
Adequate	1.00 (referent)	32.9 (1)	<001
Unclear	0.70 (0.62·0.79)		
Sequence generation	–	–	–
Adequate	1.00 (referent)	0.31 (1)	0.58
Inadequate	0.95 (0.81–1.12)		
Exclusions	–	–	–
No	1.00 (referent)	0.99 (1)	0.32
Yes	1.07 (0.94–1.21)		
Double-blinded	–	–	–
Yes	1.00 (referent)	6.16 (1)	0.01
No	0.83 (0.71–0.96)		

[a]Multiple logistic regression model with the dependent variable being binary outcome measures from each meta-analysis. The independent variables included a binary variable for treatment group (experimental vs control); indicator variables to control for the effects of each of the 229 trials; terms for the "meta-analysis by treatment group" interaction to control for the different summary odds ratios for the treatment effects in the 33 meta-analyses; and the four "quality measure by treatment" interaction terms displayed in this table to analyze their associations with estimates of treatment effects. Model deviance = 325.3; df = 192.

9.4 How to Deal With Drop-Outs

Almost all treatment studies suffer from the problem of drop-outs. It is important to notice that the number of drop-outs may affect the meaning of a study on two different levels. The first level concerns the number of patients that drop out because they do not want to continue to take the treatment or the placebo. This would mean that there is a systematic bias in the study. In drug studies, the adverse side-effects of the medication often pose a problem. In neuropsychological rehabilitation studies, the amount of time needed for treatment and the repetitive character of the training may convince patients to leave the study. If there is a systematic reason for dropping out related to the treatment itself, it may qualify only for a specific group of patients. There are several other common reasons for dropping out of a study. For example, many neuropsychological rehabilitation studies are performed at rehabilitation units, and patients stay there for a limited amount of time. They may leave the unit before the necessary number of treatments is reached. Such drop-outs indicate something about the feasibility of the treatment in specific institutional contexts but are not relevant in determining the value of the intervention itself (or in other terms, these drop-outs do not present a systematic bias concerning the goal(s) of the study). High-level RCTs document the number of drop-outs and the reasons for leaving the study in a flowchart.

How these flowcharts can be used to interpret treatment results will be demonstrated by the following two studies. Luft et al. (2008) analyzed the impact of treadmill exercise on gait performance in stroke patients. They compared the effect of treadmill exercise training for six months on gait performance compared to an active control group (The control group received the same amount of treatment, but it only involved stretching the leg muscles). They found that treadmill exercise improved treadmill walking velocity by 51% and cardiovascular fitness by 18%, and this was not the case in the control group. The authors concluded that their study shows that treadmill exercise helps to improve functional recovery in chronic stroke patients.

The study is well designed, double-blinded (the therapists offering the treatment were not blinded) and fulfills all criteria for a high-level RCT. The authors also documented a drop-out flowchart for their study (Fig. I.9.1). There were 57 patients randomized to the intervention group and 56 to the active control group. Additionally, 6 patients dropped out of the intervention group and 11 dropped out of the active control group. This is almost double the number of patients who dropped out from the control group compared to the intervention group. What might have happened? One idea is that the control participants may have noticed that their treatment did not concern their ability to walk (because they never walked during their treatment). Therefore, they might have been less motivated than the patients of the intervention group to finish the study. In this respect, the study may not have been blinded. Thus, the documentation of the drop-outs provides important additional information for the interpretation of this high-level study.

The second study to show the importance of documenting drop-outs concerns neuropsychological treatment of executive functions and memory in patients with multiple sclerosis (Fink et al., 2010). This study was not

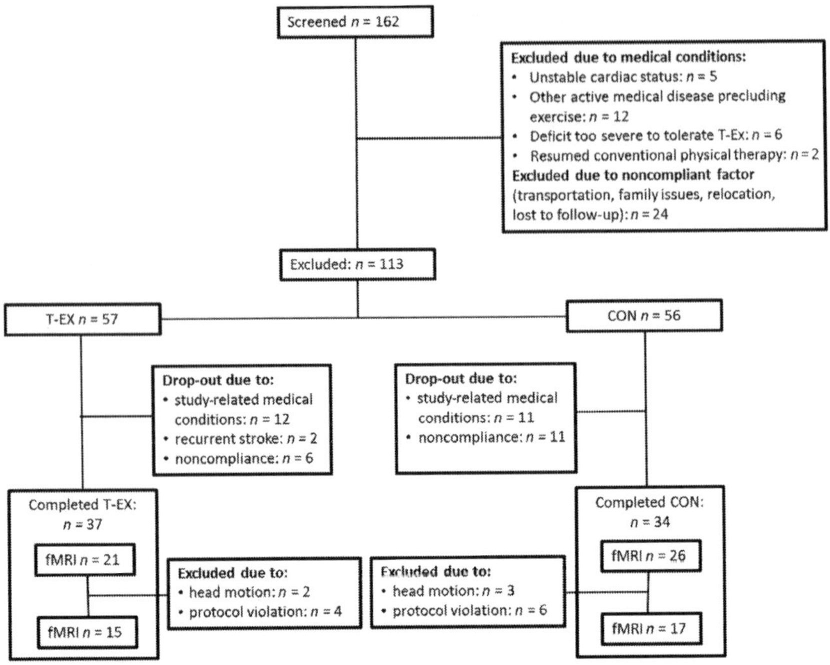

Fig. I.9.1 Participant overview (flowchart) provided by Luft et al. (2008).

completely randomized (but rather only pseudo-randomized) for reasons that are unimportant here. Three groups were included: one verum group, which received working memory training (for example, complicated calculation tasks) and training to develop plans for going shopping, to visit friends, etc. (similar to the Plan-a-day module of the RehaCom module); an active control group that trained for alertness using a professional attention training program on a computer, and finally a third untreated control group. The study shows an improvement of the verum group compared to both control groups directly after the training in both executive functions and verbal memory performance. When comparing the verum group and the active control group, this effect was still present after one year. Ignoring the methodological problems of the study (for example the pseudo-randomization): Does this study argue that patients with multiple sclerosis should be treated for their cognitive deficits? The flowchart of the study (Fig. I.9.2) gives additional information on this question.

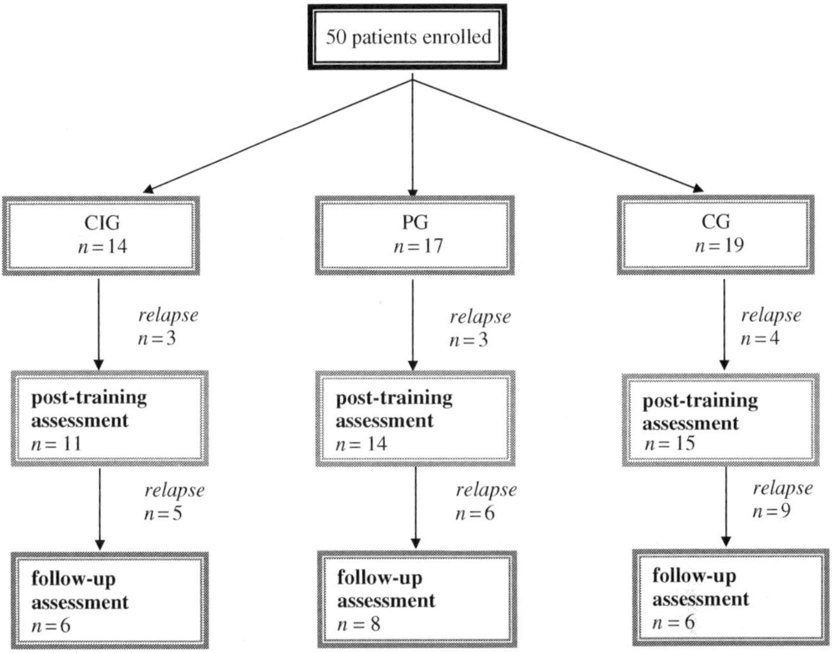

Fig. I.9.2 Flowchart of the treatment study of working memory and executive training in patients with relapsing remitting multiple sclerosis (Fink et al., 2010). *CIG*, cognitive intervention group; *PG*, active placebo group; *UG*, untreated group.

More than half of the patients suffered from a relapse during the observational time of one year and the number of patients available for the follow-up evaluation was very low. Hence, even if the results look promising, the number of drop-outs is far too high to determine whether cognitive training should be incorporated into the treatment of MS patients. First, it has to be shown that the treatment effects survive the impact of a MS relapse on cognitive processing and this question is not answered by the study.

As can be seen from the first flowchart of the Luft et al. (2008) study, this group not only documented the drop-outs during the intervention period, but also during the patient recruitment period. The flowchart shows that 162 patients were contacted and only 113 patients took part in the study. This type of information may also be important to have a complete understanding of the meaning of the study. From this we can see that the treatment was applied to a much smaller group than originally intended. Some treatment may be interesting, but because of the relatively small participation rate of the patients it cannot be implemented in health care as standard treatment.

Up to now, the strategy of dealing with drop-outs was to exclude them from further analysis. This might be envisaged as the most natural strategy, because these patients did not receive the treatment. On the other hand, drop-outs may actually boost the measured effectiveness of a treatment that would be much lower if no patients dropped out. The process of dropping out may correspond with a selection bias: only patients who experience improvement (due to whatever cause) remain in the study, and therefore the study leads to significant results. To prevent such a selection bias, another strategy is to keep the last result of any participants that drop out included in the analysis (last observation carried forward analysis, LOCF). Depending on the time-course of the treatment or the disease, this latter strategy may be either conservative or permissive in judging the treatment effect. If the symptoms are worsening during the course (for example in neurodegenerative disease) the strategy is permissive. If patients drop out because they do not experience recovery, the strategy may be conservative. In any case, if the number of drop-outs is high, the results of a LOCF analysis should be presented.

The effect of such an analysis is shown in Fig. I.9.3. It depicts the cognitive performance in patients with Alzheimer's disease who were treated with memantine measured by two different scales (Reisberg et al., 2003). On the right side of each figure one can see the outcome measured by these scales after 28 weeks. The results are shown for the endpoint using LOCF analysis (outermost right), which includes all the patients with at least one existing evaluation point, and for week 28 (last assessment) without using a LOCF analysis. The figures show that more patients dropped out of the placebo group and that the drop-out rate had an impact on the efficacy analysis. This is because by using the LOCF evaluation, a significant difference between the groups could only be shown with one of the scales. The treatment itself and the placebo condition may lead to a selection bias that was present during the study. Conducting an additional LOCF analysis may help to eliminate that bias. On the other hand, it should be clear that such an evaluation strategy is very conservative, because patients are included that did not receive the treatment anymore. In this respect, the LOCF analysis shares aspects with the ITT analysis, which was discussed earlier.

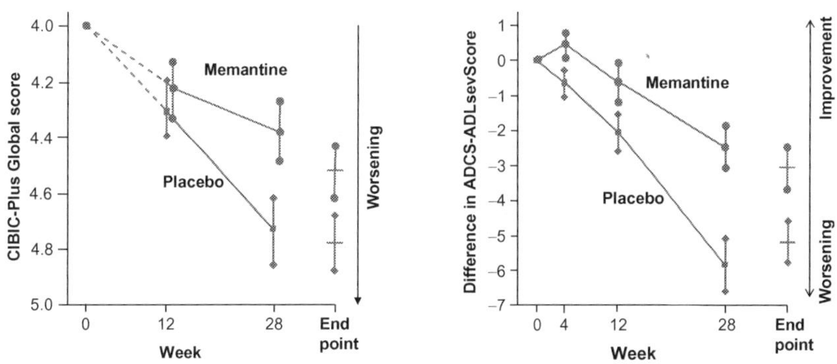

No. at risk				
Memantine	126	107	97	118
Placebo	126	105	84	118
P value			0.03	0.06

No. at risk						
Memantine	126	119	107	97	124
Placebo	126	117	106	84	123
P value				0.003	0.02

Fig. I.9.3 The impact of memantine on the progression in patients with Alzheimer's disease (Reisberg et al., 2003).

9.5 A Summary of Bias Evaluation

The Cochrane collaboration developed a systematic scoring system to estimate the different biases that have the potential to affect the results of interventional studies (an example of the results of a this scoring system is shown in Fig. I.9.4). The risk for bias can be judged as high, unclear, or low. The first two items of this scoring system concern the randomization strategy and the concealment. The next three entries concern blinding of the participant, the personnel performing the intervention, and the personnel

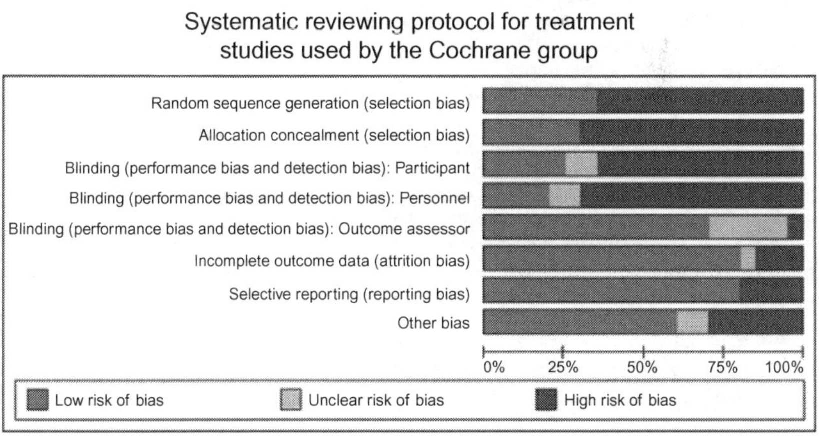

Fig. I.9.4 An example of a report that can be created when using the Cochrane review group systematic evaluation of risk scale.

assessing the outcome. The next entries concern whether there was an uneven drop-out tendency in the patient groups. The two remaining entries concern a complete reporting of the data and the impression that there might have been any other type of bias.

The PEDro scale (Fig. I.9.5) is another instrument for systematically reviewing treatment studies in neuropsychology and clinical psychology. The items of this scale are somewhat less technical than the reviewing scale of the Cochrane collaboration group and focus on additional aspects such as eligibility criteria and basic principles of statistical evaluation.

PEDro scale

1.	Eligibility criteria were specified	Yes	No	Where:
2.	Subjects were randomly allocated to groups (in a crossover study, subjects were randomly allocated an order in which treatments were received)	Yes	No	Where:
3.	The groups were similar at baseline regarding the most important prognostic indicators	Yes	No	Where:
4.	There was blinding of all therapists who administered the therapy	Yes	No	Where:
5.	There was blinding of all therapists who administered the therapy	Yes	No	Where:
6.	There was blinding of all therapists who administered the therapy	Yes	No	Where:
7.	There was blinding of all assessors who measured at least one key outcome	Yes	No	Where:
8.	Measures of at least one key outcome were obtained from more than 85% of the subjects initially allocated to groups	Yes	No	Where:
9.	All subjects for whom outcome measures were available received the treatment or control condition as allocated or, where this was not the case, data for at least one key outcome was analysed by "intention to treat"	Yes	No	Where:
10.	The results of between-group statistical comparisons are reported for at least one key outcome	Yes	No	Where:
11.	The study provides both point measures and measures of variability for at least one key outcome	Yes	No	Where:

The PEDro scale is based on the Delphi list developed by Verhagen and colleagues at the Department of Epidemiology, University of Maastricht (Verhagen et al., 1998. The list is based on "expert consensus" not, for the most part, on empirical data. Two additional items not on the Delphi list (PEDro scale items 8 and 10) have been included in the PEDro scale. As more empirical data comes to hand it may become possible to "weight" scale items so that the PEDro score reflects the importance of individual scale items.

Fig. I.9.5 Another instrument for reviewing treatment studies: the PEDro scale (Verhagen et al., 1998).

10 SOME BASIC ASPECTS OF STATISTICAL ANALYSIS OF TREATMENT STUDIES

10.1 The Special Meaning of Binary Outcome Measures

Whereas experimental results are most often expressed in continuous variables (response times, BOLD measures, etc.), treatment effects are most often expressed in binary categories (success, yes/no). There is a clear clinical advantage in using binary data in treatment research: Treatment success can be defined in the easiest way and it concerns the success for a single patient ("Number needed to treat", increase or decrease of a risk). Such figures can be calculated only by using binary data. Therefore, reporting the success of treatment studies should not only use continuous variables but also percentage of binary events (such as improved or remediated versus not improved or remediated).

In an optimal situation, the success of a treatment is defined by an external criterion. One easy example would be: Does neuropsychological rehabilitation of neglect in an intensive care unit reduce the risk of the patient being transferred to a nursing home after the end of the treatment? Another example would be whether memory rehabilitation reduces the risk of losing one's job. These are very clear-cut definitions of outcome measures and they would immediately show that the neuropsychological intervention is clinically relevant. There are two problems with these types of outcome measures in neuropsychological rehabilitation research: First, the defined event depends on many factors (not only on the neuropsychological intervention) and therefore it is hard to achieve a significant difference just by focusing on one aspect. To show an effect from one intervention may only be possible in a very large patient sample. Second, there are many more specific aspects concerning the independence of the patients that are influenced by the existence of neglect (self-care, eating, wheelchair mobility etc.). To focus only on the binary event of where the patients go after leaving the intensive care unit may miss changes in performance that are important, but not important enough to avoid being placed in a nursing home. Therefore, neuropsychological rehabilitation studies often do not have external clear-cut outcome measures, but rather scales and test results that indicate the progress the patient has experienced by receiving the treatment.

Scores derived from scales or tests can be used for nonparametric statistical testing. Due to the tradition to report binary events in EBM, they can also be converted into binary outcome figures. The simplest way to do this is to divide all patients into two groups: those who improved and those who did not. Although this is an easy and straightforward solution, there is an inherent problem in reorganizing these types of results into binary scores. For many of the test results there are no objectively defined cut-off values which determine a *clinically* relevant treatment success. To give a typical and highly debated example: During the last decade of the twentieth century, the research group of Bernhard Sabel (Kasten et al., 1998) reported that it is possible to reduce the amount of visual field loss after brain damage by approximately 2–4 degrees of visual angle by treating patients with restitution therapy (which is a very exhausting and long-lasting treatment). Before any results can be noticed, patients must train for an average of 50 h at a computer. Furthermore, there are no well-defined parameters to determine which patients benefit from the treatment and which do not. While the treatment increases the visual field significantly (in statistical terms), does an increase of about 2–4 degrees also have clinical relevance? Many colleagues in the field would say no, but the debate is still on-going. For example, in patients with no sparing of the foveal visual field, an increase of 2 degrees might mean that they can read again.

The lack of well-defined cut-off scores not only concerns test results, but it also presents a problem when using clinical scores. For most of the scores it is unknown which *differences* mark a clinically relevant change that could be used to indicate whether the treatment was efficient on a single patient level. However, some clinical scales (such as the fatigue scales) at least differentiate between mild, moderate, and severe fatigue, and a change from one level to another may be regarded as clinically relevant (Sander et al., 2016; Sander, Voelter, Schlake, Eling, & Hildebrand, 2017). Without a definition of clinical range, the amount of improvement in a given scale will be arbitrary to a certain degree. In this case, more global clinical conventions may be used. Such a convention would state, for example, that any improvement above 15% is clinically relevant. More often, cut-off values like improvements above 0.5 or 1 standard deviation are used. It should be clear that using the standard deviation as a surrogate for an external validation is a suboptimal solution.

10.2 Risk and Risk Reduction

Binary outcome data are often expressed as "risk". In this case, the "risk" to reach a binary endpoint is expressed as the percentage of patients reaching or – for a better understanding of the term "risk" – avoiding a negative endpoint. The major comparison then is the reduction of risk upon reaching the endpoint (death, persistent dependence on nursing care) due to the intervention. Binary data are usually evaluated by statistical tests such as the Chi-square test. Where confounding variables must be taken into account (such as age, severity of a disease or symptom) the evaluation can be undertaken with a logistic regression analysis. In this type of analysis, one of the independent variables is membership to the verum or the control group.

Before going into more detail on how to deal with risk and risk reduction measures in treatment studies, it should be mentioned that the primary outcome variable should be defined at the very beginning of a study. In designing treatment studies, the effect of running multiple tests on the chance of yielding a significant result is of major importance. As neuropsychological rehabilitation studies usually include a relatively small number of patients, the p-value for the treatment effect will not be as large as in drug studies. If the p-value must be divided through a number of statistical tests, the chance of concluding with a significant result is not very high. Defining one (or not more than two) primary endpoints means that all the other tests are tests of secondary endpoints, and these tests do not have to be adjusted for multiple testing. The other side of the coin is that missing significant results in the primary endpoints means that the study has failed, even though the results concerning the secondary endpoints may be fine or even excellent.

As already mentioned, there are some conventions for how to report binary outcome measures, which are not all intuitively evident (Table I.10.1). The easiest report concerns the absolute risk reduction: The risk of an event in the control group minus the risk of an event in the treated group, which is usually expressed as a percentage. In the neurological rehabilitation unit of the Klinikum Bremen-Ost, two-thirds of the patients suffering from a brain infarct or bleeding reach the next level of rehabilitation. Introducing a new and more effective treatment strategy would reduce the risk of not reaching the next level. The efficacy of the new treatment can be reported by calculating the risk reduction: The percentage not reaching the next level (33.3%) minus the percentage not reaching the next level, out of those getting the new treatment (e.g., we will say 20%). The absolute risk reduction of the new treatment then would be 33.3%–20% = 13.3%.

A somewhat less-intuitive measure is the relative risk reduction of the new treatment. This treatment led to an absolute risk reduction of 13.3%. The relative risk reduction is defined as the division of the absolute risk reduction by the absolute risk of the control group, i.e. 13.3%/33.3% = 40%. In other words, the new intervention removed 40% of the risk to not reach the next rehabilitation level from the original risk.

One may also compare the risks of both groups directly: The risk for the standard treatment group is 33.3%, and the risk for the group receiving the new treatment is 20%. Therefore, the relative risk of the treatment group compared to the standard group is 20%/33.3% = 0.6.

There odds ratio is another possibility for the comparison of the two groups. For the standard treatment group, the event of not reaching the next

Table I.10.1 Conventions for how to report outcome measures

Measure of effect	Abbreviation	Description	No effect	Total success
Absolute risk reduction	ARR	Absolute change in risk: the risk of an event in the control group minus the risk of an event in the treated group; usually expressed as a percentage	ARR = 0%	ARR = initial risk
Relative risk reduction	RRR	Proportion of the risk removed by treatment: the absolute risk reduction divided by the initial risk in the control group; usually expressed as a percentage	RRR = 0%	RRR = 100%
Relative risk	RR	The risk of an event in the treated group divided by the risk of an event in the control group; usually expressed as a decimal proportion, sometime as a percentage	RR = 1 or RR = 100%	RR = 0
Odds ratio	OR	Odds of an event in the treated group divided by the odds of an event in the control group; usually expressed as a decimal proportion	OR = 1	OR = 0
Number needed to treat	NNT	Number of patients who need to be treated to prevent one event; this is the reciprocal of the absolute risk reduction (when expressed as a decimal fraction); it is usually rounded to a whole number	NNT = ∞	NNT = 1/initial risk

level of rehabilitation is given nominated as 1 and the event of reaching it is nominated a 2 (or in other words, two out of three patients are reaching the next level and one is not). In the new treatment group, the ratio is 1 to 4: Four patients reach the next level and one patient does not. The odds ratio then is ¼ / ½=0.5. Although the odds ratio is widely used in EBM, it is not easy to intuitively understand it.

The last value (and possibly most important from a patient's perspective) is the number of patients that must be treated with the new treatment to prevent one event. This number (also called number needed to treat) is defined by the reciprocal of the absolute risk reduction, and it is usually expressed as a decimal fraction which is then rounded to a whole number. In our case, the number needed to treat is 1/0.13=7.7. Therefore, about 8 patients must be treated with the new treatment so that we can be sure that one patient (that would have otherwise failed to progress) moved to the next rehabilitation level.

Table I.10.2 shows another example of these different parameters. This table shows an example (not a result of a real study) for a neuropsychological rehabilitation study on neglect rehabilitation. There was one binary outcome parameter for this hypothetical study: reaching the next rehabilitation level or leaving for home care. One hundred and sixty patients were randomized either to treatment-as-usual or to treatment-as-usual plus neuropsychological rehabilitation of neglect. The intervention was successful: Only 37 of the 80 patients receiving neglect rehabilitation did not reach the next rehabilitation level, whereas in the treatment-as-usual group

Table I.10.2 A hypothetical example for the efficacy of neglect rehabilitation on outcome of early rehabilitation efforts

	Care home	Next rehabilitation level	Total
Specific neglect treatment	33	47	80
No neglect treatment	48	32	80
Total	85	75	160

Calculation of risk, risk reduction etc. for Table I.10.2.
Risk care home without treatment: 48 × 100/80= 60%.
Risk care home with treatment: 33 × 100/80= 41.25%.
Absolute risk reduction: 60% − 41.25%=18.75%.
Relative risk: 41.25/60%=0.69.
Relative risk reduction: 18.75%/60% × 100 = 31.25%.
Odds ratio: 33/47/48/32=33 × 32/47 × 48 = 0.46.
Number needed to treat: 1 × 100/18.75 = 5.3.

48 failed in this respect. Please take some time to calculate the different percentages and ratios, which were defined above.

Cabrera-Gomez et al. (2003) analyzed the impact of intramuscular injection of Interferon beta 1a on cognitive decline in patients with Multiple Sclerosis over a period of 104 weeks. For this purpose, the performance of the patients was assessed by the PASAT (Paced Auditory Serial Addition Test, which measures performance in working memory and attention), which is part of the Multiple Sclerosis Functional Composite Score. Fig. I.10.1 shows the diagram provided by the drug company to document that Interferon beta 1a reduces the risk of progression in cognitive decline. For this purpose, 0.5 standard deviation was used as a cut-off score and patients crossing this difference were counted as additionally impaired. After 102 weeks, 19.5% of the treated patients deteriorated, compared to 36.6% of the placebo group. This meant an absolute risk reduction of 17.1% and a relative risk reduction of 17.1% /36.6% = 47%.

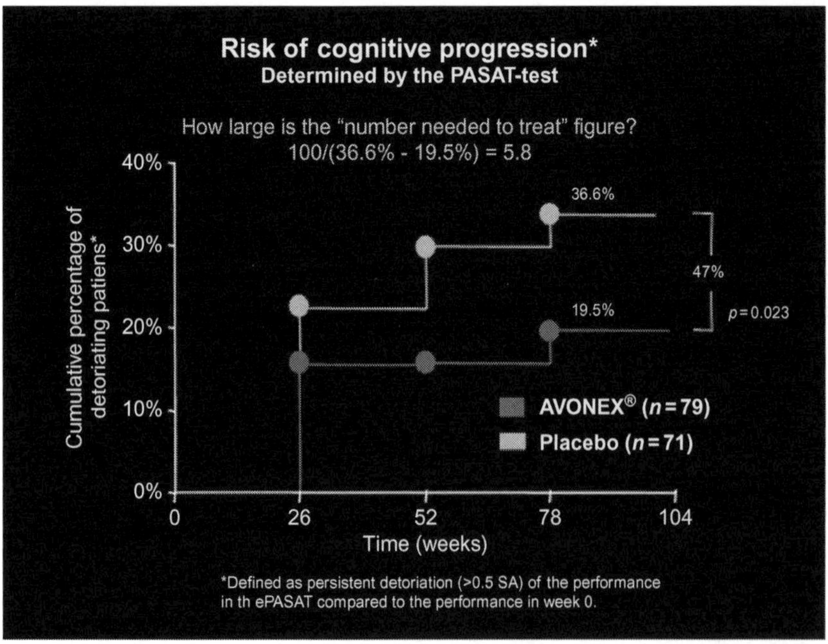

Fig. I.10.1 The impact of Interferon beta1a on cognitive deterioration (Cabrera-Gomez et al., 2003).

10.3 Clinical Scales as Outcome Measures or How to Deal With Ordinal Data

If no simple binary outcome measures are available, or the expectation to show an impact on such figures is low (for example because they are also influenced by various other causes), broadly accepted clinical scales can be used to show a relevant clinical outcome of the intervention. As already mentioned, the BI and the FIM are potential scales that could be used, but there are also more specific ones for neuropsychological rehabilitation. Examples include the Catherine Bergego Scale for Neglect, the "Fragebogen erlebter Defizite der Aufmerksamkeit" (FEDA) for attention, the Cognitive Failure Questionnaire for executive functions, etc. Clinical scales use scaled scores for judging the performance of patients. The FIM, for example, allows for a classification of memory performance starting with "1", i.e. no memory performance, to "7", i.e. completely normal memory performance. As the distances between the single numbers for the grading of memory performances differ, the data are nonparametric data. This holds for all clinical scales.

Ordinal data are analyzed by nonparametric statistical testing, which is defined and explained in any textbook on statistical testing. For rehabilitation studies, it should be noted that the main burden to show the efficacy of a method relies on the comparison of the groups at *t2*. If there are significant differences, the intervention is argued to be effective. Nonsignificant differences at t1 are ignored as "chance" differences. Due to correct randomization, no differences at t1 are expected. Therefore, not all studies report the results of the comparison of the groups at t1. Intragroup improvements, which are often reported, do not really qualify as markers for treatment efficacy as long as there is no difference at t2 between the groups. For directly comparing the treatment effects in different groups the results of t1 and t2 have to be subtracted. This is because nonparametric testing does not allow for a simultaneous testing of group and time differences.

10.4 Parametric Evaluation of Treatment Results

There are two different options to analyze treatment effects with multivariate parametric tests. The first option is to use analysis of variance (ANOVA) for repeated measurement or analysis of covariance (ANCOVA) for repeated measurement. In the latter case the group status is the independent variable, the two or more assessment points are the dependent variables, and one may add possible confounding variables as covariates. There should be an interaction effect between the independent and the dependent variables showing that the verum groups show a larger improvement than the control group. Please note that both groups should also differ in the post hoc *t*-Test for the post interventional assessment. Fig. I.10.2 shows a typical example of such a statistical analysis: the main effects of the variable group and of the intervention (irrespective of group status) are analyzed first, and then the interaction between these two variables are calculated. A positive effect of the intervention is indicated by a significant interaction effect.

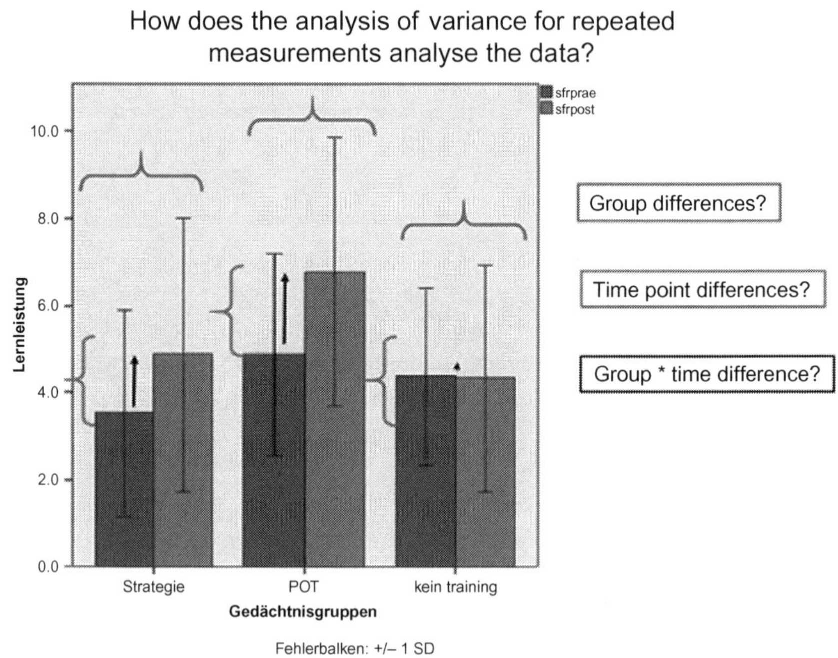

Fig. I.10.2 An example of how to analyze treatment data by an analysis of variance for repeated measurement.

The other possibility to test intervention effects by parametric statistical analysis is regression analysis. Regression analyses are difficult to interpret and partly dependent on the method (inclusion, forward,). The group status (verum versus control group) will be one of the independent variables, while the treatment effect will be the dependent variable. In the case of a regression analysis, the treatment effect should be coded binary (yes or no) to use logistic regression analysis. This is not completely necessary; one may also subtract the assessment results and use this difference as the dependent variable in a linear regression analysis. However, in logistic regression analysis one is able to specify whether the independent variables are categorical or not (e.g., gender). Because the advantages of a regression analysis are that one may specify confounding variables and see if the treatment effect is independent of these variables, it is helpful to be able to use such categorical variables. The main question is whether the group status remains in the regression analysis, which argues that the intervention was effective.

10.5 Power Analysis for Designing Treatment Studies

As has been argued in the paragraphs above, undertaking neuropsychological rehabilitation research is time consuming and complicated. It is also risky, because negative findings in RCTs will be reviewed by the Cochrane collaboration and will have an influence on the guidelines for neurological rehabilitation. Therefore, doing research in neuropsychological rehabilitation should rely on a reasonable risk calculation about the prospect of the study (this naturally holds true also for treatment studies outside the domain of neuropsychology). If one believes that an intervention might help patients with acquired brain damage, the most important question is how large the effect will be that intervention would produce. This depends on the primary outcome parameter one would like to influence. The second related question is, what does this mean for the number of patients that have to be treated to show a significant difference in the primary outcome parameter? Obviously, this number depends on the efficacy of the treatment. If the treatment is very good and the rate of spontaneous recovery in the control group is zero, only one patient has to be treated with this method to show that it works. On the other hand, if the spontaneous recovery rate is 50% then the two patients in both groups may show recovery. Therefore, one has to calculate how many patients (with such a spontaneous recovery rate) have to be compared with the same number of treated patients to find a statistically significant difference between the both groups. As interventions are often not helpful for all patients, and because there is always some spontaneous recovery or placebo effect in untreated patients, a calculation of the number needed to be included in the verum and the control group is necessary to avoid negative results. The opposite is also true: Because neuropsychological rehabilitation is time consuming, it is interesting to know the lowest number of patients where one may expect significant results.

The statistical knowledge domain for these types of questions is called power calculation (for an introduction, see Kutschmann, Bender,

Grouven, & Berg, 2006a). Power calculation is one additional criterion of a high-level treatment study, because it helps the reviewer ensure that the positive or negative result of the study is based on a reasonable number of patients. Otherwise the results may just present a publication bias, i.e. a positive but chance finding. Power calculation relies on two different questions. The first one asks: How many patients have to be treated to yield a significant difference between the groups? This is the usual question in neuropsychological experimenting and a significant difference is usually expressed by a p-value of <0.05. The second question is: How sure do I want to be that I really found a difference of $p < 0.05$? It is possible to miss this cut-off score just by including some patients that differ a little bit from normal patients or by deviating just a little bit from what was originally intended to be done with the intervention. This is a type II error: Although the intervention makes a difference, there is a risk in not finding this difference with a given number of included patients. As in the case of the type I error (the statistical test shows a $p < 0.05$ significant difference, but there remains the (<5%) chance that this difference is irrelevant), there is a conventional cut-off score for type II errors. This cut-off score is 80% and it means that in 20% (or fewer) of the investigations the difference will be missed, even though it exists.

Changing these two cut-off scores will influence the number of patients that have to be included in the study (as well as the effect size that is expected for the intervention). Reducing the chance for a false-negative result means that one must include more patients to avoid or compensate for any arbitrary negative variations. The same holds for looking for a p-value smaller than 0.05. Fig. I.10.3 shows the influence of the effect size of an intervention on the number of patients that must be included to yield a significant result of $p < 0.05$ and a power of 80%. The diagram also shows that this number depends on the standard deviation of the primary outcome measure before the intervention starts. If this standard deviation is high, many more patients must be recruited compared to if the standard deviation was low. To aid in understanding this concept, one may think of a measure that is very noisy. Depending on how noisy the measure is, one needs more (or fewer) measurement trials to be sure that there is really a difference between the two groups. A similar graph could be constructed varying the p-value and the intended power.

At the same time, the relation between the standard deviation and the number of patients to be included in a study showing the effectiveness of an intervention poses a problem for neuropsychological rehabilitation studies. It implies that the original distribution of the primary outcome parameter is known in order to calculate the necessary number of patients. This is often not the case for neuropsychological deficits, which vary according to

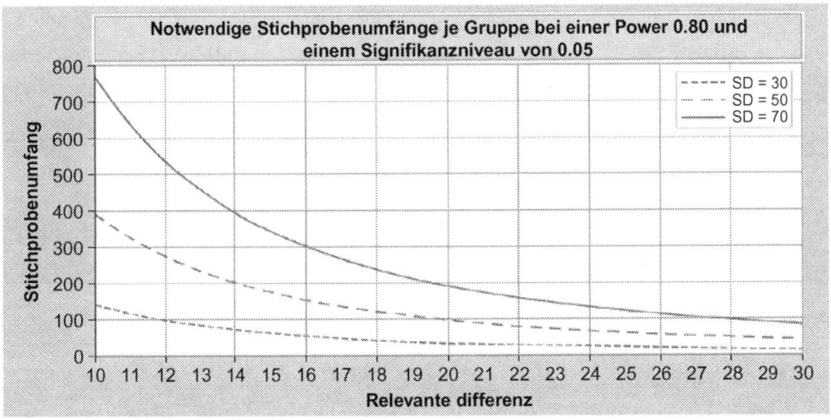

Fig. I.10.3 Number of patients to be included (*y*-axis) according to the estimated effect of the intervention (*x*-axis) and the variability of the measurement (legend). *(From Kutschmann, M., Bender, R., Grouven, U., & Berg, G. (2006). [Aspects of sample size determination and power calculation illustrated on examples from rehabilitation research].* Rehabilitation (Stuttg), 45*(6), 377–384. https://doi.org/10.1055/s-2006-940113.)*

the heterogeneous composition of patient groups, and which develop differently depending on the time since lesion (acute, subacute, and chronic). Neglect, for example, means something quite different in an early rehabilitation unit as opposed to in a Phase D rehabilitation unit. Therefore, it is often not possible to rely on already-given distributions for specific deficits when trying to calculate the necessary power of a study. There is no simple solution to this problem, but one possibility is to assume a similar standard deviation for both groups (verum and control group). This is reasonable because of the random allocation of the patients. In this case, it is quite easy to calculate the number of necessary patients for different assumptions about the efficacy of the intervention (see below).

The extended formula to calculate the number of patients for a treatment study is shown in Fig. I.10.4.

$$n = \frac{\sigma_{mN}{}^2+\sigma_{mN}{}^2}{\mu_{mN}-\mu_{oN}} \cdot (z_{1-\alpha/2}+z_{1-\beta})^2 = \frac{70^2+70^2}{21^2} \cdot (1.96+0.84)^2 = 174$$

σ : Standard deviation of the distribution

μ: Mean

mN, *oN*: with and without intervention

$Z_{1\text{-}\alpha/2}$ = Normal distributions value of $p = 0.05$ => 1.96

$Z_{1\text{-}\beta}$ = Normal distributions value of $\hat{a} = 20$ => 0.84

Fig. I.10.4 Power calculation for continuous variables (Kutschmann et al., 2006a)

In this formula, n is the number of patients needed. The two variables above the division line represent the square of the standard deviations of the verum and the control group. The two variables below the division line represent the difference between the mean in the outcome parameter; the estimated effect of the intervention. The two z-scores depict the values for a $p < 0.05$ (or less) and a power of 80% (or less). The term on the right side of the equal sign is an example for two groups with a standard deviation of 70 and a difference in their means (induced by the intervention) of 21 multiplied with the sum of the z-values for a $p < 0.05$ and a power of 80. In this case, the number of patients that must be included is 174. Due to the necessary two-sided t-test, this number has to be doubled ($n = 374$).

This formula can be reduced when p is always <0.05, the power is always 80%, and the standard deviations are always the same. As already mentioned, this is reasonable because of the intended randomization of the patients (Fig. I.10.5).

$$n = 16 \cdot \frac{\sigma^2}{d^2} = 16 \cdot \frac{4900}{441} = 178$$

σ: Standard deviation of the distribution
d: Estimated difference for outcome between the intervention and the control group

Fig. I.10.5 Simplified power calculation for continuous variables. *(From Kutschmann, M., Bender, R., Grouven, U., & Berg, G. (2006). [Aspects of sample size determination and power calculation illustrated on examples from rehabilitation research].* Rehabilitation (Stuttg), *45(6), 377–384. https://doi.org/10.1055/s-2006-940113.)*

This reduction of the formula is possible by adding 1.96 + 0.84, calculating its square value (~ 8) and multiplying the results with 2 because of the two identical standard deviations above the division line. This yields the 16 which has to be multiplied with the division of the square of the standard deviation by the square of the estimated difference between the group means after the intervention. Rewritten in this way, the formula allows for the calculation of the number of patients to be included in a very simple manner: If the estimated effect size of the intervention is one standard deviation, the division in the formula is 1 and the number of patients is 16 × 2 (due to two-sided testing). If the estimated effect size is 1/2 of a SD the number is 64 × 2, in the case of 1/3 of the SD, 144 × 2 etc. In other terms, the number of patients needed for recruitment can be standardized on the relation between the effect size of the intervention and the standard deviation: $(16 \times 2) \times x^2$, where x is the division of the SD by the expected effect size.

Concerning the multiplication with 2, it may be argued that there is a clear hypothesis: Only the intervention will change the results and therefore two-sided testing is not necessary. In my view, this argument is correct, but there might be reviewers that argue that just waiting might also have an effect, because the placebo effect in the control group might be larger than the (detrimental) effect of the intervention.

So far, we have dealt with continuous variables. However, as already mentioned, neuropsychological rehabilitation research should also focus on binary outcome data like "recovered: yes or no". In this case the calculation of significant differences is done by Chi square statistics and looks a little bit more complicated at first glance (Fig. I.10.6).

$$n = \frac{\left(z_{1-\alpha/2}\sqrt{2\pi_{\emptyset}(1-\pi_{\emptyset})} + z_{1-\beta}\sqrt{\pi_1(1-\pi_1)+\pi_2(1-\pi_2)}\right)}{(\pi_1-\pi_2)^2}, \quad \pi_{\emptyset} = \frac{(\pi_1+\pi_2)}{2}$$

$$n = \frac{(1.96\sqrt{2\cdot 0.0625\cdot 0.9375}+0.84\sqrt{0.05\cdot 0.96+0.075\cdot 0.924})2}{(0.075-0.05)2} = \frac{0.918}{0.000625} = 1469$$

$\pi 1$ = healthy without intervention = 5 %

$\pi 2$ = healthy with intervention = 7.5 %

$\pi_{\emptyset}$ = ($\pi 1$ + $\pi 2$) / 2 = 6.25 %

Fig. I.10.6 Power calculation for binary variables.

Again, we have the two *z*-scores for the $p < 0.05$ and the 80% power of the study. The other terms are now percentages of patients in a group. In the presented case, it is assumed that without the intervention 5% of the people will recover spontaneously, whereas with the intervention 7.5% will recover. The average chance to recover for both groups is 6.25%. Because the difference between the number of recovered patients in both groups is small, the number of patients to be recruited to reveal a significant effect with $p < 0.05$ is high. Fortunately, this formula can be reduced to a simpler one (Fig. I.10.7).

$$n = 16 \cdot \frac{\pi_{\emptyset}(1-\pi_{\emptyset})}{d2} = 16 \cdot \frac{0.0625 \cdot 0.9375}{0.000625} = 1500$$

Fig. I.10.7 Simplified power calculation for binary variables. *(From Kutschmann, M., Bender, R., Grouven, U., & Berg, G. (2006). [Aspects of sample size determination and power calculation illustrated on examples from rehabilitation research].* Rehabilitation (Stuttg), 45*(6), 377–384. https://doi.org/10.1055/s-2006-940113.)*

In other words: To find the number of patients to be recruited in a study it is necessary to multiply the risk to recover for the whole group with the risk for the whole group not to recover. This is then divided by the square of the difference in the risk to recover between the two groups. The result is then multiplied by 16.

Let us take the example about the reduction of the risk to fail in reaching the next rehabilitation step by neuropsychological neglect therapy (which is presented above, Table I.10.2), using the equation:

$$n = 16 \times ((48/80 + 37/80)/2) \times (1 - (48/80 + 37/80)/2)/(48/80 - 37/80)^2$$

$$n = 16 \times (0.6 + 0.4625)/2 \times (1 - (0.6 + 0.4625)/2)/0.1375^2$$

$$n = 16 \times (0.53125 \times 0.46875)/0.01890625$$

$$n = 859.45 \text{ or } n = 860$$

Thus, in this example, 860 patients must be treated (remember that Chi-squares tests are one-sided) to show a power of 80% or a statistical difference with $p < 0.05$ that the treatment works.

Initially, this result is very disappointing because it implies that nobody will do such a study. This is because it is very hard to recruit so many participants and to then treat them for several weeks.

On the other hand, the difference of 11 patients between the groups is small: just above 1/5 of the patients improved according to the chosen outcome parameter. Assuming a difference of a whole standard deviation means that 2/3 of the population have moved above the cut-off score. This would mean in the example that the number of patients not reaching the next rehabilitation level is reduced from 48 to 16 patients (i.e. of 2/3).

In this case the formula can be solved in this way:

$$n = 16 \times ((48/80 + 16/80)/2) \times (1 - (48/80 + 16/80)/2)/(48/80 - 16/80)^2$$

$$n = 16 \times (0.6 + 0.2)/2 \times (1 - (0.6 + 0.2/2)/0.4^2$$

$$n = 16 \times (0.4^* \, 0.6)/0.16$$

$$n = 24$$

This example shows again that the number of patients to be included in a treatment study to yield a significant difference heavily depends on the effect size of the intervention.

Because neuropsychological rehabilitation research will be based on relatively small patient numbers, (as it is very time consuming), intervention studies will have to rely on clever and effective ideas to show significant results.

11 PROBLEMS WITH GROUP STUDIES IN NEUROPSYCHOLOGICAL REHABILITATION RESEARCH

Without a doubt, group studies, especially RCTs, are the most convincing instrument to show that an intervention is helpful in improving the outcome of patients. No other method or research design can control for bias and other methodological problems on such a high level as RCTs. However, there are some inherent problems with RCTs when using this methodology in neuropsychological rehabilitation research. The first and most important problem is that double blinding (or in rehabilitation research, triple blinding) is very hard to achieve. It is not clear what placebo means in rehabilitation research. Most patients will realize that "just doing something with them" is not a real treatment. The same holds for the person providing the intervention. Therefore, the realization of triple blinding in neuropsychological rehabilitation research requires the comparison of real interventions. This can be accomplished either with different intensive schedules or with different content. However, note that most of the recent drug studies on antidepressants would have shown negative results if compared to an active comparator, i.e., with a real treatment.

Providing a placebo treatment is also problematic from an ethical viewpoint. In contrast to drug studies, patients are actively participating in rehabilitation studies. It is therefore hard to tell them that the many hours they spent under the assumption of receiving treatment were all for nothing. This is extremely problematic if the study is done in a rehabilitation setting and the placebo treatment is *instead* of getting a real treatment. In this case, the patients will likely not get any aimed treatment at all. On the other hand, doing neuropsychological rehabilitation research with chronically ill patients (i.e. outside a rehab unit) may yield negative results, whereas the same treatment might be effective during the first weeks after the acquirement of brain damage. There are many studies showing that the brain shows an increased capacity to reorganize early after suffering from brain damage. Moreover, it is much more complicated to set up a neuropsychological rehabilitation study outside of a rehab unit. This is because the patients must come to the building where the treatment takes place, not all patients will be able to drive alone, and only highly-motivated patients will be prepared to come five times a week for several weeks. Such a group of patients might imply a selection bias. The study on treating MS patients for executive and memory deficits (Fink et al., 2010), which was already mentioned before, had to be performed as a pseudo-randomized study because of the driving time to the hospital. Only patients from the neighborhood were willing to attend the meetings at the hospital, but these meetings were necessary to provide the treatment.

When patients must drive (or be driven) to the treatment location, it implies that they should be reimbursed for their travel costs. In contrast to drug research, there is no significant funding for research on neuropsychological rehabilitation. Finally, achieving the necessary case number is complicated due to the heterogeneity of the patients and the large amount of time required to recruit a homogenous patient group. We tried to investigate the efficacy of a specific training program for patients' isolated memory impairments following a stroke of the posterior cerebral artery (PCA). The areas of the brain often affected after this type of stroke are the hippocampus and/or the thalamus. Such patients will probably show different memory deficits than patients with memory impairments where the hippocampus is not involved. But PCA infarcts account only for 5% of all infarcts. Additionally, half of them occur in the right hemisphere without substantial verbal memory impairment. Even after a left hemisphere PCA infarct in some of the patients only the visual cortex is affected, i.e. not the first segments of the PCA, which supply the hippocampus and/or the thalamus. It took three years to achieve a three-armed study, with 15 patients each, in a Phase D rehab unit with 100 beds. Unfortunately, the results were negative (at least on a statistical level). Nevertheless, looking at the data of individual patients, there was a small number with some high effects (e.g., Fig. I.11.1). Therefore, group studies may also overlook important relations that become visible on a single case level.

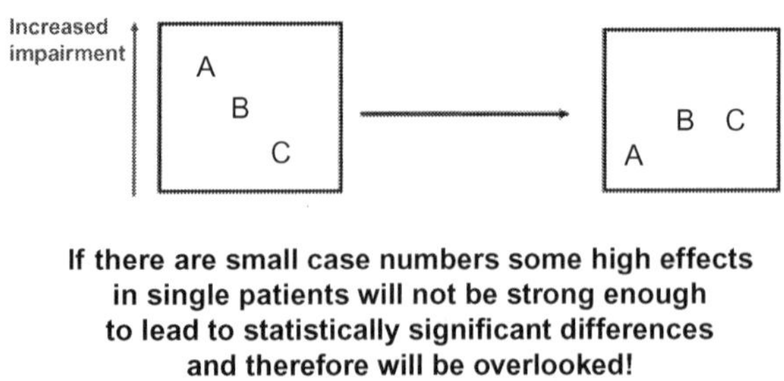

Fig. I.11.1 Statistical averaging may ignore important changes in individual patients.

On this basis, some colleagues have argued that it is best to stick to single cases and to ignore group studies. For a variety of reasons, this seems to be the wrong solution to a legitimate problem. Nevertheless, it should be evident that well-designed single case studies can be extremely helpful in a situation where group studies pose a major problem. For example, they can be useful in instances where the specific symptoms to be treated are rare (e.g., pure alexia), or where a group study would be very laborious and the expectations about the treatment effect are still unclear. Group studies and SCEDs have their problems and their merits. In the next section, therefore, the methodology of the SCED for treatment studies will be discussed.

12 SINGLE-CASE EXPERIMENTAL DESIGNS IN REHABILITATION RESEARCH

12.1 Introduction and Definition

EBM relies mainly on group studies, but clinical decisions are based on the observations of single patients who have received certain interventions. The "most natural" type of evaluation of whether an intervention is effective, therefore, is the pre- and post-treatment comparison. In clinical routine, the evidence of certain interventions (based on group effects and evaluated on the basis of specific inclusion criteria) is not always clear in single cases (e.g., effectiveness of treatments developed for patients with pure neglect in patients with neglect and hemianopia, in patients with memory disorder and global impairment, or cognitive slowing versus those with only memory disorder, etc.). In the best clinical routine practice, there is a need to adapt different interventions to single cases. To a certain extent, the diagnostic process spent with a patient is also a type of single case analysis (e.g., which type of cueing might be helpful to the specific patient with aphasia undergoing treatment: phonological or semantic cueing?). Evidence from group and single-case studies are complementary: The adaptation to single cases often generates hypotheses for the next group study. As shown in the previous section, because neuropsychological rehabilitation research is a very time-consuming process, it is very reasonable to investigate the effects of interventions first in single cases and then in group studies.

B. A. Wilson (1987) tried to argue that SCED treatment studies are even more rigorous than group studies and that neuropsychological rehabilitation should focus on SCEDs more or less exclusively. In the time since its publication, the methodology of SCED studies has been improved. Nevertheless, it should be evident that SCEDs will not be able to replace group studies and the arguments for this will be discussed in a later section. Single cases are also somewhat suggestive because positive treatment evidence will be perceived as more meaningful than negative evidence. Therefore, there is a certain risk that the researcher (and therapist) will be biased in the wrong direction.

Before going into more detail on the problems of SCEDs, let us first turn to the question of what defines a high standard SCED (Tate, Perdices, McDonald, Togher, & Rosenkoetter, 2014). SCEDs can be defined as a repeated measurement of a single person in different phases of intervention and nonintervention in a specific sequential structure. Therefore, any SCED includes at least one (longer) phase of baseline measurement and the phase of intervention. The main concept behind SCEDs is that the patient functions as their own control.

There are also some criteria for high level SCEDs: First, there should be an inherent and logical relation between the intervention and the behavioral change. For example, memory rehabilitation should lead primarily to changes in memory performance and not, for example, in neglect. Second, the intervention should target behavioral change primarily in one respect. This change in behavior should become evident after the intervention. Therefore, the behavioral change should be measured repeatedly and should be reproducible.

This latter criterion is very important but has the potential to be problematic. Imagine you would conduct a SCED to study physical problems. The causality between the intervention and the result should be directly visible, and it should disappear if you stop your intervention. It is only when both criteria are fulfilled that you can assume the full proof of causality.

The methodological background for a SCED is similar. It always starts with the observation of what the behavior is like without the intervention (the baseline or "A" phase). Next there is the intervention, and the defined outcome parameter should show a difference (the intervention or "B" phase). Afterwards the behavior should go back to the original state (the second baseline or "A"). A SCED therefore always has an A–B–A structure.

Only observing the baseline and then the results of the intervention is an incomplete paradigm, and it will not prove that the behavioral change is dependent on the specific intervention. It may be the endogenous course of the disease or the symptom that explains the "success" of the intervention (which is usually distributed over multiple sessions).

Although the A–B–A structure of a SCED is mandatory, this does not mean that it is the only methodological design that can be used (Tate et al., 2014). The first variation is to introduce different ("multiple") baselines and to measure the behavior during these baselines. The A–B–A structure of the SCED predicts that the behavior will change only in response to the specific treatment, i.e. in the sequence of the introduction of the different interventions. If the behavioral improvement can only be observed in cases where the target of the intervention was that specific behavior, and all other behavior is not affected by the intervention, than this is a strong argument for a causal link between treatment and success. Furthermore, this also goes against the assumption that it is only a higher global motivation resulting from the intervention that is responsible for the changes.

Another possibility is to try different interventions to change the same behavioral deficit, either with exactly two different interventions ("crossover design" or A–B–A–C–A design) or with even more than two interventions ("multi treatment design"). The frequency of the interventions can also be increased, i.e. A–B–A–C–A–C–A–B–A). One may also use an alternating treatment design (A–B–C–B–C–A), but this design has some methodological problems. This is because there may be a carryover effect of one intervention to the other, which cannot be measured without the existence of baselines.

SCED may also be used to test dose effects on behavior. In this case the number of treatment sessions defining the B phase are systematically increased (A–B–B–A → A–B–B–B–A), and the assumption is that the amount of behavioral change (which can be measured) is contingent on the number of B sessions.

12.2 Some Typical Examples and Some Typical Problems

Fig. I.12.1 is a prototypical drawing of a SCED: It starts with three baseline assessments, moves on to an intervention period, and involves two baseline assessments at the end of the cycle. There should be an improvement of performance after starting with the intervention. After finishing the intervention, the behavior goes back to its original state.

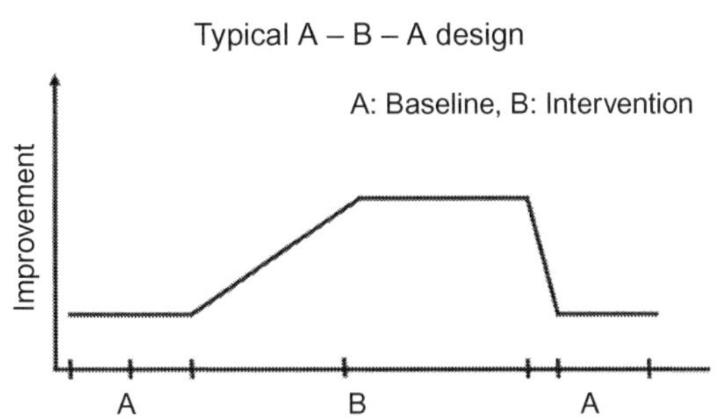

Fig. I.12.1 An example of a SCED in the A–B–A design.

In her paper on SCED as a method for neuropsychological rehabilitation research, B. A. Wilson (1987) documented the consequences of introducing a timer on the ability of a patient with severe amnesia to remember a new name (Fig. I.12.2). The timer helped the patient to retrieve the name just before it vanished from memory. Wilson showed that the timer helped the patient to remember the name for several (up to 10) minutes. Any longer period without the external signal to try to remember the name resulted in the patient forgetting it. When the timer was removed, the period in which the patient could remember the name went back to just one minute. When the timer was reintroduced, the period of time it was possible to remember the name went up again.

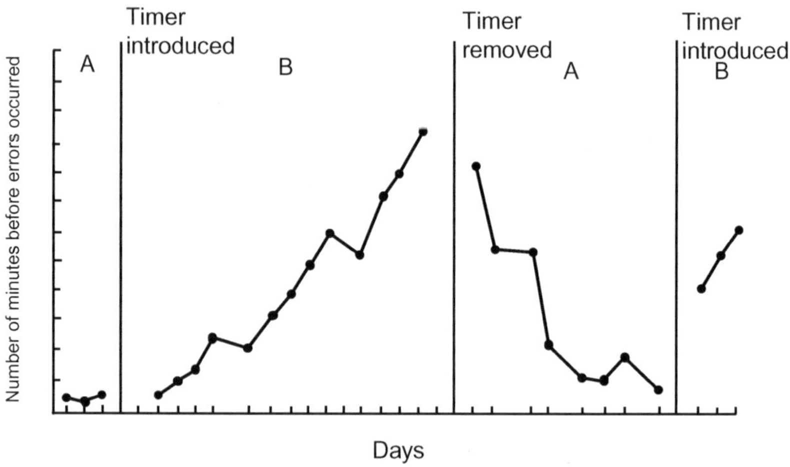

A single-case ABAB design. Evaluating the effect of a timer to recall a name after intervals of gradually increasing length

Fig. I.12.2 Effects of a timer on remembering a name displayed as a SCED graph (B. Wilson, 1987).

At least two things are remarkable for this SCED. First, it focused on a chronic case and did not induce a lasting change in overall memory performance. Rather, it focused only on remembering one specific piece of information. Second, the performance was completely dependent on the intervention, i.e. the introduction of the timer and an externally driven, spaced retrieval schedule. Although this is clearly a possible scenario for neuropsychological rehabilitation, it is not the only possibility. In many cases, the goal of the intervention is to change the functional performance *permanently*, not only for the period of the intervention and for one specific item. While in severely and chronically impaired patients a permanent change in performance may be impossible, it might be a reasonable goal in only moderately impaired and post-acute patients.

Working with the assumption that the behavior is not completely dependent on the intervention but may be changed permanently through relearning or reorganization violates the methodological assumptions of SCED. If withdrawing the intervention does not reinstate baseline behavior again the second proof for causality (B → A) is lost. Consequently, SCED suffers from specific problems which will be discussed in relation to the following case study.

Kerkhoff et al. (2011) published a paper on the impact of galvanic vestibular stimulation on tactile extinction, in which two SCEDs were used. Galvanic vestibular stimulation is a variant of transcranial direct current stimulation. It is assumed that the vestibular nerve is affected by this stimulation and, consequently, it induces an activation of the parietal and insular cortex as well. The two patients suffered from a small brain lesion and showed extinction for tactile stimulation on the contralesional side. The effect of the stimulation was measured for different stimuli and for identical stimuli and stimulation was applied to both hands of the patients. In both patients, a baseline measure was first used, followed by a sham galvanic vestibular stimulation. Next, a right-sided galvanic vestibular stimulation was presented under the assumption that this stimulation would affect the left hemisphere of the brain. This was followed by a left-sided vestibular stimulation and after several weeks a follow-up measurement was conducted. The second patient had several more baseline measurements as well as some additional follow-up sessions. Fig. I.12.3 shows the results of the stimulation in the second patient with the large number of assessments.

Kerkhoff et al. (2011) concluded from this series of measurements that galvanic vestibular stimulation leads to a decrease of tactile extinction. They were able to show that this is true even when they used a sham control condition.

This paper is a typical paper for which the ultimate goal is not only to show that an intervention is effective, but that it is effective in changing behavior *permanently*. In such studies, the second part of the causality proof (the consequence of withdrawing the intervention) is not part of the hypothesis. Therefore, the stability of the baseline is a very important aspect of the study. This is because only a stable baseline performance and a notable change in performance starting with the interventions would be able to substitute a lack of evidence in the withdrawal part of the SCED. From a

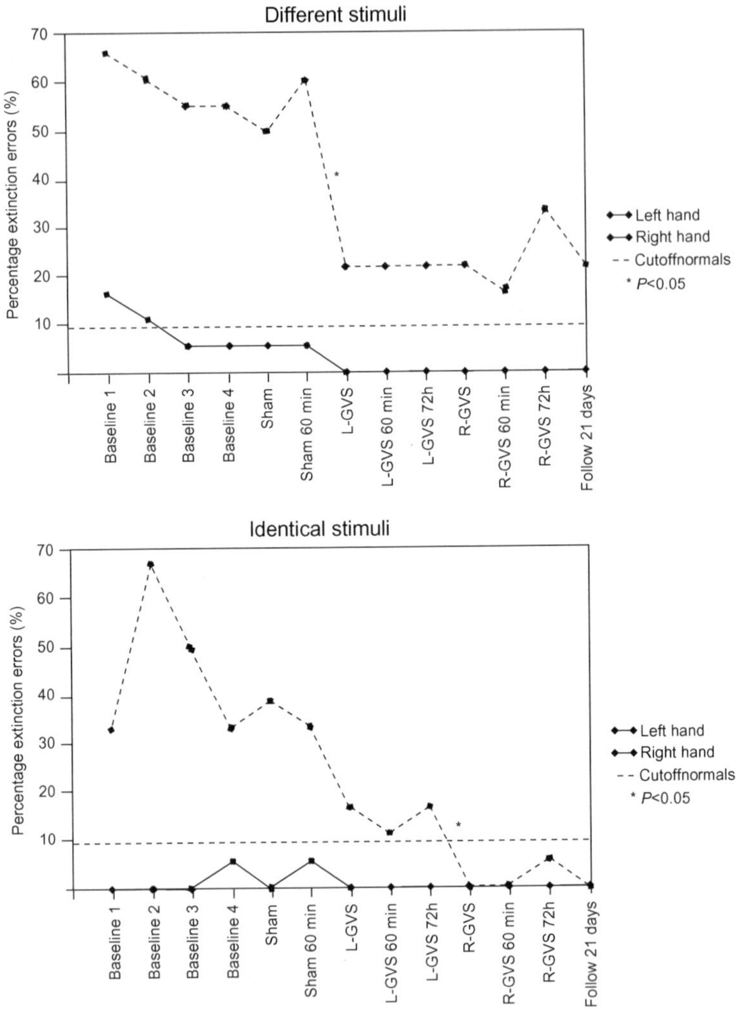

Fig. I.12.3 The effects of galvanic stimulation on tactile discrimination displayed as a SCED graph (Kerkhoff et al., 2011).

theoretical standpoint there still remains the problem of proving the causality of the intervention. However, from a pragmatic standpoint the stability of the baseline may be accepted as substitution for the lack of a rebound of the baseline performance.

How important the stability of the baseline is for the evaluation of SCEDs can be easily shown by redrawing the figures of the Kerkhoff et al. (2011) study with a plausible trendline for the development of the behavior (Fig. I.12.4).

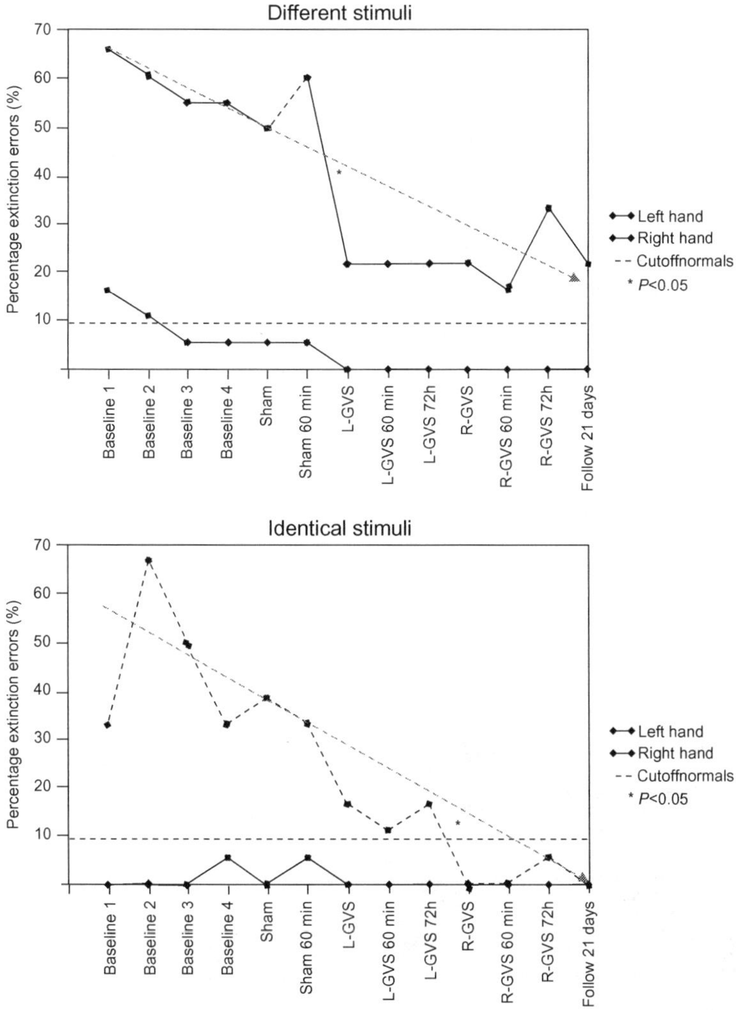

Fig. I.12.4 The importance of a stable baseline for using SCEDs in rehabilitation research: the same figures redrawn with a trend line.

As can be seen in the figures, the success of the intervention may have been a result of measuring the same behavior several times and not a consequence of the galvanic vestibular stimulation. It can be shown from that example how suggestive positive treatment results are for people providing the intervention. It also shows how important it is for SCEDs (that do not expect a reinstatement of the original behavior after withdrawing the intervention) to demonstrate the stability of the baseline or a clear difference in behavior when one compares the baseline and the intervention.

Raymer et al. (2007) reported another series of SCEDs focusing on the improvement of word-finding performance in aphasic patients. Their question was: Does cueing with semantic and phonological features of words provide an effective method to improve word finding for nouns, verbs, or for both? They also compared the effect of the method on practiced and not practiced words. They treated eight aphasic patients. Out of eight patients, five patients demonstrated an improvement. There was no difference in effectiveness of the method for nouns and verbs. Word finding of nonpracticed words did not improve at all from the cueing method, and there was no generalization from nouns to verbs or vice versa. Fig. I.12.5

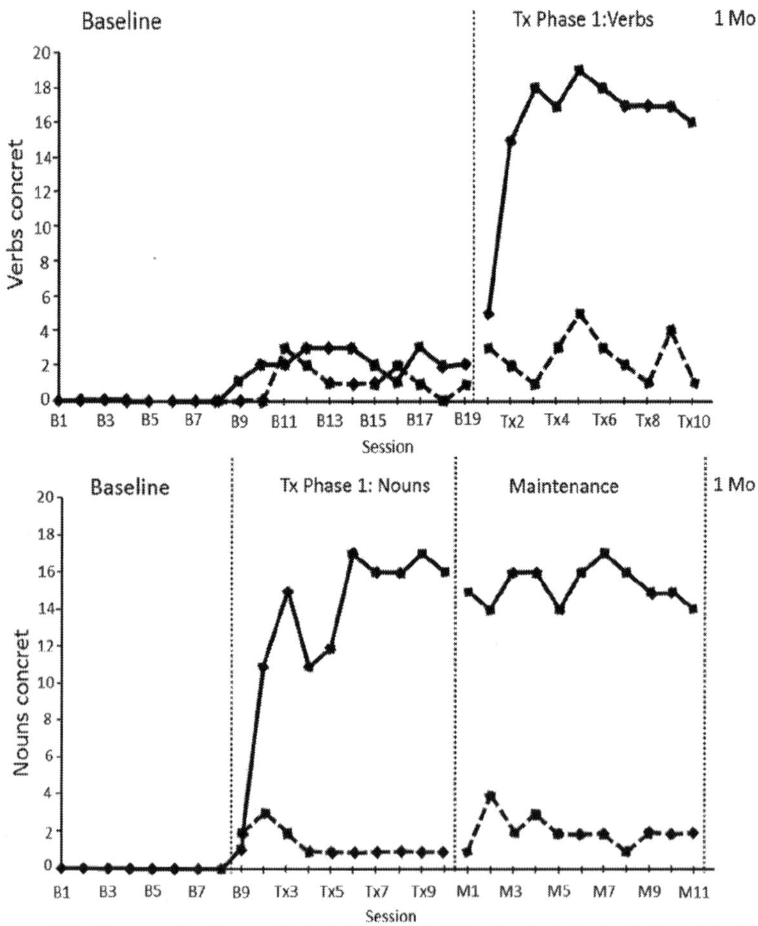

Patient 1 accuracy in picture naming for trained verbs (solid lines) and untrained verbs (dashed lines) (upper figure) and trained nouns (solid lines) and untrained nouns (dashed lines) (lower figure)

Fig. I.12.5 Effect of cueing on naming performance in two aphasic patients (Raymer et al., 2007).

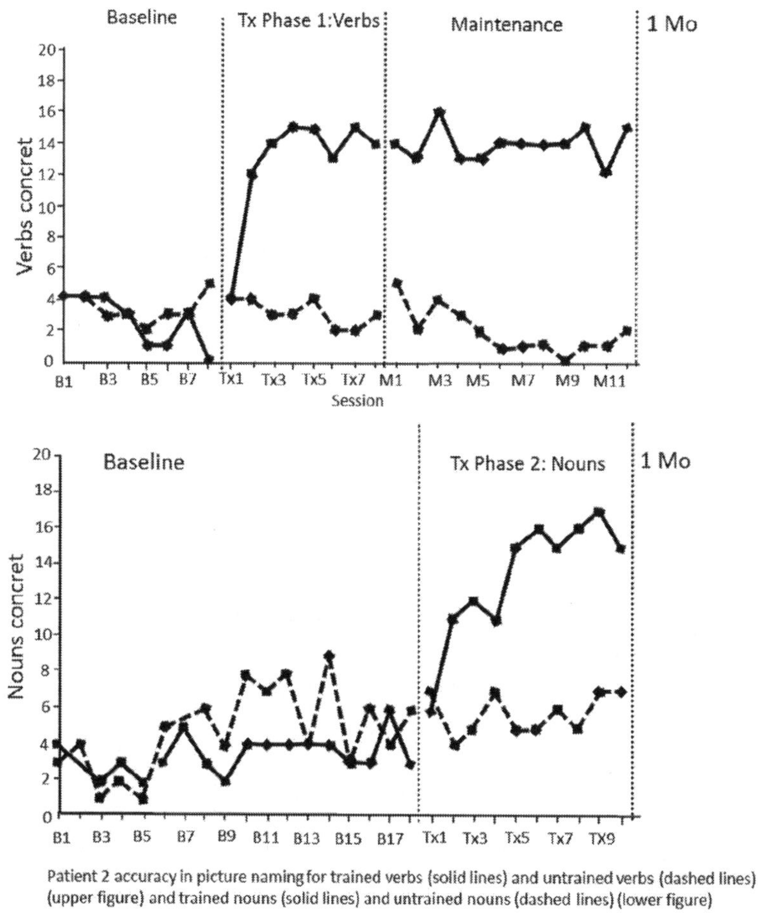

Fig. I.12.5, Cont'd

shows the results for two patients. As can be seen, Raymer et al. (2007) used a multiple baseline design and randomized the intervention sequence (starting with nouns or with verbs). The figures show an impressive increase in performance, which cannot be explained by any baseline instability. In these respects, the study is convincing.

On the other hand, only five of eight patients improved. Additionally, Fig. I.12.5 shows the greatest improvements from the study. We do not know whether Raymer et al. (2007) treated more than the eight reported patients. It is theoretically possible that some patients went unreported. Thus, in SCEDs there is the potential for a huge reporting bias problem that can be even higher than in group studies.

12.3 Improving the Methodological Quality of SCEDs: The RoBINT Scale

Many of the problems surrounding group studies and RCTs also apply to SCEDs. For example, blinding to eliminate the effect of placebo treatment is a problem common to group and SCED studies. Castro Caldas et al. (2017) reviewed different treatment studies in patients with multisystem atrophy (MSA), an atypical Parkinson's syndrome. Fig. I.12.6 shows the effect of the studies depending on the methodology used. "Parallel" stands for RCT and single arm for A–B studies (typically not blinded). The figure shows the large effect of blinding and randomization on the observed treatment effect.

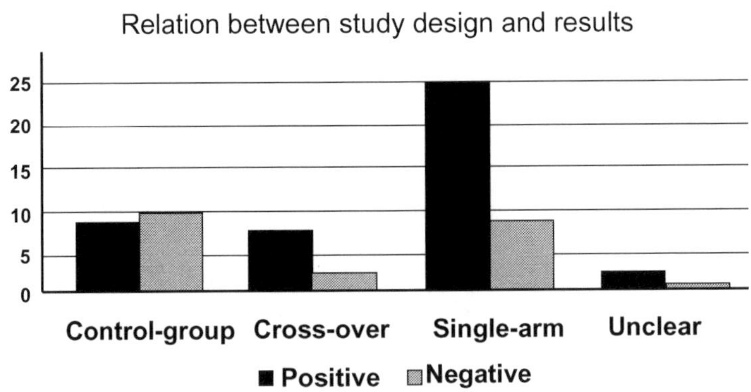

Fig. I.12.6 Relations between methodology and results (Figure taken from Castro Caldas et al., 2017).

Randomization also concerns both SCEDs and group studies. While group studies involve the randomization of patients, SCEDs require the randomization of treatment sequences. In 2014 an influential paper (Tate et al., 2014) tried to develop something like a PEDro scale for SCEDs: the risk of bias in n-of-1 trials (RoBINT scale). In the first step, the authors of the paper defined which designs are SCEDs and which are not. They classified the designs of withdrawal, reversal multiple baseline, alternating treatment, and changing criterion designs as SCEDs. All designs with only one or two phases were excluded from classification as an SCED.

They then continued to define specific qualitative features of SCEDs: the design, the randomization, the sampling behavior, the blinding of the patient/therapist/assessor, the inter-rater reliability of the assessment, the

Items of the RoBiNT Scale

Internal validity subscale

1. Design
2. Randomisation
3. Sampling behavior (all phases)
4. Blinding patient/therapist
5. Blinding assessors
6. Inter-rater reliability
7. Treatment adherence

External validity and interpretation subscale

8. Baseline characteristics
9. Therapeutic setting
10. Dependent variable (target behavior)
11. Independent variable (intervention)
12. Raw data record
13. Data analysis
14. Replication
15. Generalization

Fig. I.12.7 Items in the RoBiNT scale (Tate et al., 2014).

treatment adherence, questions of the definition of the outcome parameter, replication, and generalization (see Fig. I.12.7).

Many of these characteristics for high-level SCEDs are quite similar to those of RCTs. However, there are some methodological problems that are specific to SCEDs. Therefore, from a pure methodological standpoint it is correct to argue that the RCT is the gold standard for EBM (which does not mean that SCEDs are not also necessary and extremely helpful). The first problem is a pragmatic one and it concerns the lack of stability of the baseline in post-acute rehabilitation. It will be very hard to find clear-cut criteria on what counts as effective intervention and what as spontaneous recovery if the behavior does not go back to baseline. In post-acute interventions, spontaneous recovery will be the rule and not the exception. If this is the case, the second part of the causality proof is lacking and the surrogate proof (stability of the baseline) cannot be assumed. The solution for this, as previously discussed, is to calculate the trend for the baseline and to compare this trend with the development during the intervention. However, this is problematic when there are only a small number of baseline measurements (e.g., three or four), which is typical of most studies.

The second problem is that the intervention in an SCED is the result of a social interaction between one therapist and one patient. This raises the question of the reproducibility of the intervention in any other patient–therapist coupling, as it is known that the social interaction itself influences the effect of an intervention. Therefore, generalizability to other patients and other therapists is an open problem in SCEDs. Last, but clearly not least, there is the "world record problem" with a corresponding publication bias: It is always possible that there is one patient in 100 or 1000 patients who greatly improves for arbitrary reasons and whose case is later published. It is foreseeable that patients that show the most improvement (even if under chance or atypical circumstances) would appeal the most to researchers and journals.

SCEDs and group studies should go hand in hand in neuropsychological rehabilitation research. While group studies are very work intensive and are often underpowered because of the number of included patients, SCEDs alone are not strong enough to show treatment effects in the most convincing manner. From such a perspective, efforts to improve SCED quality such as the RoBINT scale should not lead to a situation where the flexibility and creativity of SCEDs is lost, because the methodological affordances are too high for clinical routine practice. Clearly, SCEDs should be well done, but one should keep in mind that they have mainly an important heuristic and inspiring function for rehabilitation research.

13 THE QUALITY OF A TREATMENT STUDY DOES NOT ONLY DEPEND ON A SOUND METHODOLOGY

So far, we have reviewed important methodological features of high level RCTs and SCED trials and related these designs to the defined hierarchy of EBM. However, the methodological design is not the only criterion that decides if a treatment study is accepted by the professionals working in the field. To illustrate this, let us discuss the following two treatment studies. Kober et al. (2015) published a paper on EEG based neurofeedback training and memory functions in post-stroke victims. Box I.13.1 shows the abstract of the study.

To give another example: Huang, Lai, and Tang (2007) published a paper on the effects of needling acupoints DU20, DU26, and HT7 in intervening in patients with vascular dementia, i.e., to increase their cognitive performance. Box I.13.2 shows the abstract of their study.

There are several obvious objections, which may be raised against both studies: They are completely underpowered (<11 patients per group), and

BOX I.13.1 Abstract from Kober et al. (2015).

Background: Using EEG based neurofeedback (NF), the activity of the brain is modulated directly and, therefore, the cortical substrates of cognitive functions themselves. In the present study, we investigated the ability of stroke patients to control their own brain activity via NF and evaluated specific effects of different NF protocols on cognition, in particular recovery of memory.

Methods: $N = 17$ stroke patients received up to ten sessions of either SMR ($N = 11$, 12–15 Hz) or Upper Alpha ($N = 6$, e.g. 10–12 Hz) NF training. $N = 7$ stroke patients received treatment as usual as control condition. Furthermore, $N = 40$ healthy controls performed NF training as well. To evaluate the NF training outcome, a test battery assessing different cognitive functions was performed before and after NF training.

Results: About 70% of both patients and controls achieved distinct gains in NF performance leading to improvements in verbal short- and long-term memory, independent of the used NF protocol. The SMR patient group showed specific improvements in visuospatial short-term memory performance, whereas the Upper Alpha patient group specifically improved their working memory performance. NF training effects were even stronger than effects of traditional cognitive training methods in stroke patients. NF training showed no effects on other cognitive functions than memory.

Conclusions: Post-stroke victims with memory deficits could benefit from NF training as much as healthy controls. The used NF training protocols (SMR, Upper Alpha) had specific as well as unspecific effects on memory. Hence, NF might offer an effective cognitive rehabilitation tool improving memory deficits of stroke survivors.

there is no information about blinding, concealment and drop-outs (etc.) available. Furthermore, irrespective of these methodological failures, both studies also suffer from conceptual problems. The first study defines its goal as the improvement of memory functions. Memory performance relies on limbic structures which are not related to upper alpha and sensorimotor rhythm (SMR). Memory performance may also rely on prefrontal areas, but again there is no theoretical link between the two rhythms that were trained and prefrontal memory retrieval functions. SMR is also not related to visuospatial working memory (in contrast to verbal working memory). Results that concern memory (and not working memory) performance are not reported at all. Thus, the results of the study are not anchored in any kind of theoretical model.

BOX I.13.2 Abstract from Huang et al. (2007).

Objective: Using methods of clinical scale assessment and cerebral functional imaging to compare the relative specificity of needling acupoints Baihui (DU20), Shuigou (DU26), and Shenmen (HT7) in intervening vascular dementia (VD) in different areas in the brain.

Methods: Fifty patients with VD were randomized into 5 groups. Needling on conventionally used acupoints of hand and foot three Yang-meridians aiming at hemiplegia was applied to the patients in Group A, and needling on DU20 to Group B, on DU26 to Group C, on HT7 to Group D and on all the three to Group E was applied additionally. Assessments of Mini Mental State Examination (MMSE), Activities of Daily Living (ADL) and Family Attitude Questionnaire (FAQ) were made. And the positron emission computerized tomography (PET) and single photon emission computerized tomography (SPECT) examinations were conducted in 5 selected patients from each group before and after treatment.

Results: Needling on conventional acupoints plus DU20 could effect the inner temporal system, thalamencephalon system, and prefrontal cortical system to improve memory and executive capacity of VD patients; conventional acupoints plus DU26 could effect more to the prefrontal cortical system to obviously elevate the executive capacity; that plus HT7 would reveal an effect similar to but rather weaker than plus DU20, and effect more to memory; and that plus all the three simultaneously could effect rather roundly multiple aspects of the nervous system related to intellectual activities, to elevate the recognition and enhance the executive capacity.

This becomes even more obvious in the second study. Currently, there is no evidence at all as to why acupuncture of the hand, foot, or face should differentially increase brain activity. One might imagine that acupuncture influences the body (restricting specific movements, etc.), but Huang et al. (2007) argue that there is a specific and enduring effect on different brain areas.

Progress in treatment is based on progress in understanding body functions. This has been true in the development of modern medicine. The same holds for progress in neuropsychological rehabilitation. Progress in neuropsychological rehabilitation depends on progress in understanding the functional architecture of cognitive functions. It is on this basis that specific treatments may be developed, and the next step may be refined after being successful. The quality of a study does not only depend on its methodology, it also depends on its theoretical clarity and elegance, i.e. on its foundation in and its motivation by cognitive models. The concept of theoretical plausibility and an anchoring in background models are not part of EBM like the evaluation of a study. However, they are a presupposition for a steady progress, where treatment studies build on each other and verify models that are developed in experimental research. The plausibility of the relation between the kind of intervention and its effects also plays a major role in the tacit evaluation of studies' results by the experts of the field.

We will try to develop this conceptual aspect of rehabilitation research in the field of the treatment of memory disorders a little bit further without ignoring the importance of methodological soundness.

CHAPTER II

Rehabilitation of Moderately to Mildly Impaired Memory Functions

1 FREQUENCY (PREVALENCE) AND PERSISTENCE OF MEMORY DISORDERS AS CONSEQUENCE OF ORGANIC BRAIN DISEASES

1.1 Prevalence

Memory disorders are common in neurological diseases. A study on cognitive deficits in newly diagnosed Parkinson's disease patients showed that at this early point of the disease 8.4% already suffered from a memory deficit. In multiple sclerosis patients, memory impairment is present in about 40% of all patients with relapsing remitting course and even higher in patients with secondary chronic course. In patients with traumatic brain injury (TBI), herpes encephalitis or hypoxia, memory impairments are very often consequences of their diseases.

Snaphaan and de Leeuw (2007) published a review of studies on the prevalence of post-stroke memory dysfunction in nondemented patients (Table II.1.1). They found 65 articles on this topic. The prevalence of memory impairment varied between 13% and 50% in the first days after the stroke. The large difference can be explained partly by different cut-off scores for memory impairment (two standard deviations versus below the 10th percentile) and partly by the different time-points at which the stroke patients were assessed.

Several years ago, we investigated the frequency of memory impairment in patients admitted to a rehabilitation unit specialized for the C and D level of neurological rehabilitation (Müller, Hildebrandt, & Sachsenheimer, 2000). About 750 patients were included in this study (Table II.1.2, Fig. II.1.1) and assessed for attentional, visuoconstructional, visual field, and memory deficits. Using cut-off scores of the California Verbal Learning Test (CVLT), about 60% of the patients had retrieval deficits and about 46% had learning deficits. Additionally, 62% of the patients suffered from an ischemic stroke or an intracranial bleeding, about 7% from nontraumatic subarachnoidal bleeding,

Cognitive Rehabilitation of Memory
https://doi.org/10.1016/B978-0-12-816981-0.00002-3
© 2019 Elsevier Inc.
All rights reserved.

Table II.1.1 Prevalence of post-stroke memory dysfunction in nondemented patients in different studies (Snaphaan & de Leeuw, 2007)

Single memory assessment; Reference*	Stroke survivors				Prevalence of poststroke memory dysfunction							
						Stroke-memory assessment interval in months						
	No.	Mean age, y (SD)	First-ever stroke	Memory test	Cut-off	0.25	1	3	6	7	12	>12
4	16	59 (NS)	Y	AVLT	<10th percentile	50%						
37	78	NS	NS	RBMT	Score failed seven or more items					49%		
38	220	59 (13)	NS	WMS	<2SD			23%				
35	30	76 (8)	NS	WMS	<1SD		50%					
36	229	56 (11)	Y	AVLT	<15th percentile			31%				
Multiple memory assessment; reference												
18	196	69 (13)	Y	AVLT	<10th percentile		24%		17%		15%	
5	139	69 (12)	Y	AVLT	<10th percentile		22%		12%			
16	60	61 (NS)	Y	WMS	NS			55%			31%	
19	23	56 (16)	Y	AVLT	<2SD	13%						11%
17	111	60 (14)	Y	AVLT	*Z*-Score < 1.65	22%				6%		

Note: Empty cells indicate no assessment.
NS indicates not specified; Y, yes; AVLT, Auditory Verbal Learning Test; RBMT, Rivermead Behavioral Memory Test; WMS, Wechsler Memory Scale.
*Reference numbers according to original publication in Snaphaan and de Leeuw (2007).

Table II.1.2 Neurological patients evaluated for calculating the need of neuropsychological rehabilitation in a phase C and phase D rehabilitation center (Müller et al., 2000)

Age distribution	Female	Male	Total
Age < 20 years	25	25	50
Age 21–30 Years	20	29	49
Age 31–40 Years	29	42	71
Age 41–50 Years	45	65	110
Age 51–60 Years	59	115	174
Age > 60 Years	133	160	293
Total	312	436	748
Age (in years)	57.65 (SD 16.61)	53.91 (SD 15.26)	55.43 (SD 15.95)
Duration of illness (in months)	18.29 (SD 50.37)	10.59 (SD 39.12)	14.42 (SD 45.46)

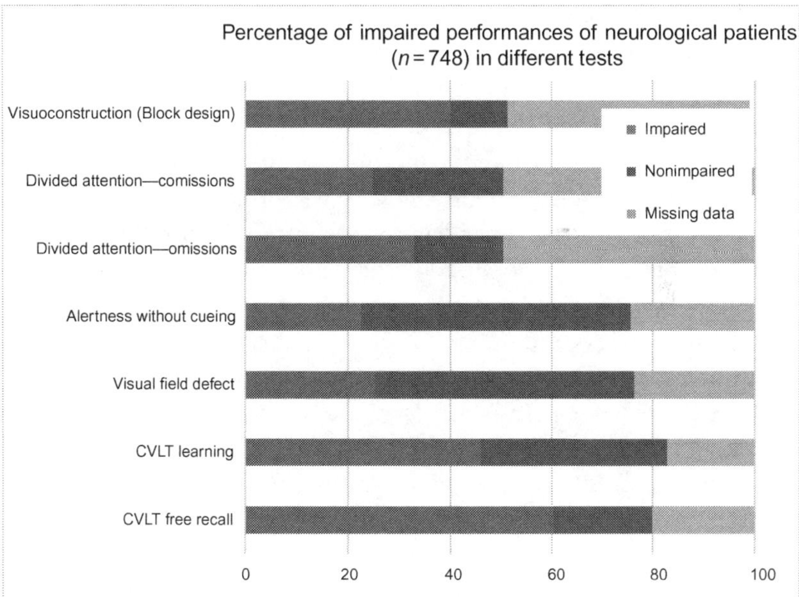

Fig. II.1.1 Prevalence of cognitive impairments in the phase C and phase D neurological rehabilitation center (Müller et al., 2000). Impairment is defined by a performance below the 16th percentile.

and 11% from a TBI. Most of the patients showed a mild to moderate memory impairment and not a dense amnesia. This is a slightly higher prevalence than in the Snaphaan and de Leeuw (2007) study. This may be explained by the fact that we only included patients who were admitted to a neurological rehabilitation unit, while Snaphaan and de Leeuw (2007) included all patients who suffered from any type of stroke, but excluded demented patients (who will show memory deficits per definition). Moreover, our sample was composed not only of stroke patients but also by those who suffered from global hypoxia, from TBI, and from encephalitis, and these patient groups have a higher prevalence of memory disorders than stroke patients.

We can conclude from these data that memory dysfunctions are frequent in neurological diseases, with an incidence of about 30% (depending on the etiology) in the acute phase and at the hospital, but above 50% in rehabilitation units. Therefore, rehabilitation of memory deficits is an important task in neurorehabilitation.

1.2 Recovery, Persistence of Memory Disorders, and the Possibility for Treatment

Spontaneous recovery from memory deficits after acquired brain damage has not been studied extensively. The figures of the review by Snaphaan and de Leeuw (2007) show that after a stroke some recovery occurs in around one-third of the patients. The amount of recovery will depend on the size and the localization of the lesion that causes the memory deficit. In HM, the famous US patient that developed anterograde amnesia after a surgical resection of the mediobasal temporal lobes, no recovery was noticed over the course of multiple decades. On the other hand, many patients show distinguished learning deficits or even anterograde amnesia immediately after a TBI, which disappears a few days after the event. Animal research provides the possibility to investigate the amount of recovery under more controlled conditions. Hildebrandt (2017) and Hildebrandt, Lehmann, and Kastrup (2012) recently reviewed the persistence of memory disorders in animals after stroke, TBI, and focal resection of brain tissue. The possibility of recovery was directly proportional to the lesion location: In the case of middle cerebral artery strokes there was an almost complete recovery if the animals (mostly rats) were provided with a stimulating environment ("enriched environment"). Rats that were kept in simple cages without the possibility of running and exploring the environment did not show recovery from memory dysfunctions. It should be noted that in rats (as in humans) the middle cerebral artery does not supply the hippocampus or other structures of the Papez circuit with blood. On the other hand, as the review of Snaphaan and de Leeuw (2007) shows, memory disorders in humans remain present for a considerable number of patients even suffering from middle cerebral artery infarcts.

In animals with artificial TBI, the recovery from memory dysfunctions was similar to stroke recovery, but it was somewhat slower. The chance for recovery was low to absent if the lesion concerned the hippocampus or the Papez circuit directly, either by surgical lesioning or by TBI. After focal and bilateral lesions of the essential structures for memory, recovery could only be noted after many weeks of enriched environment placement (if at all). Moreover, recovery was limited to easy memory tasks. Tasks that are developed to test episodic memory-like processes in animals, such as the Morris Water Maze test, did not show improvements.

In some of the animal studies, treatment effects of enriched environment could also be documented on a biological level, reducing beta amyloid load and stabilizing newly generated cells in the choroid plexus that migrated into the surrounding brain tissue (Hildebrandt, 2017). Thus, one lesson of the animal studies is that memory rehabilitation may even interact with neurobiological processes of reorganization on a cellular level and is not restricted in the sense of just using already existing cognitive tricks to compensate for the impairment.

Summing up these lessons from animal research, it seems fair to state that there is a good chance of recovery from memory deficits if animals suffer from unspecific lesions that do not damage the limbic structures per se and if animals are kept in an enriched environment (which might be compared to rehabilitative efforts in humans). Animals with unilateral lesions of the limbic system also show some recovery, but it is delayed compared to animals with nonfocal lesions. Moreover, bilateral lesions of the limbic system seem to lead to long-lasting impairments that can be treated only partially (if at all).

The review of animal research on the recovery of memory performance after suffering brain damage conforms with the clinical impression of memory-impaired patients. The German guideline for memory rehabilitation (Thöne-Otto, 2012) (Fig. II.1.2) distinguishes between post-acute and chronic patients, as well as between patients with moderate to mild memory impairment and those with dense anterograde amnesia. For both patient groups, memory rehabilitation should look for the possibility to introduce external memory aids and memory organizers to reduce the functional load experienced by the patients. For patients in a post-acute stage where the prognosis of the memory impairment is still unclear, and for patients in a chronic stage with only moderate to mild memory impairments, the

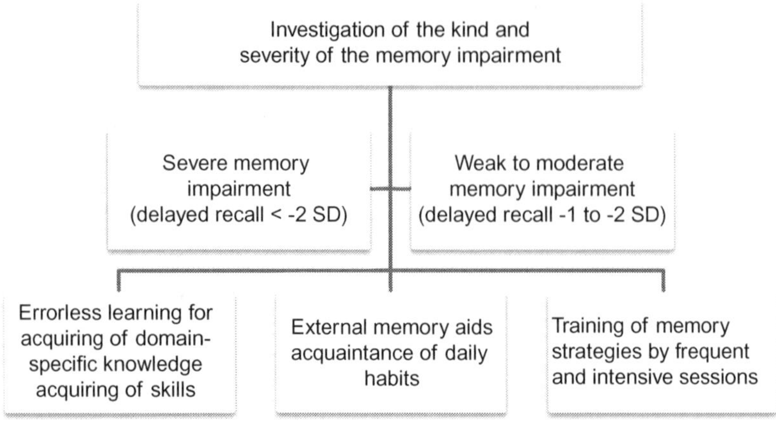

Fig. II.1.2 Flowchart for treating memory impairments from the German Guideline *Diagnostik und Therapie von Gedächtnisstörungen* (Thöne-Otto, 2012).

guideline proposes to try compensatory training concentrating on visual imagery or semantic structuring.

In the following sections, we will first concentrate on patients with low to moderate levels of memory impairment, followed by a look at patients with complete anterograde amnesia.

2 REHABILITATION OF MEMORY: IMPROVING ENCODING THROUGH MNEMONIC STRATEGIES

2.1 Memory Rehabilitation Via Teaching the Patients Mnemonic Strategies

Research on memory rehabilitation started in the 1980s. A good starting point in the literature from this time is the book by Barbara Wilson entitled *Rehabilitation of Memory* (1987), in which she presented her seminal studies on this topic. Although the book is very skeptical of the prospect of memory rehabilitation and strongly biased in the direction of behavioral therapy, Wilson conducted a series of studies investigating the efficacy of mnemonic strategies in improving the performance of memory-impaired patients. Her cognitive model for memory impairment and rehabilitation was the Atkinson and Shiffrin (1968) model (Fig. II.2.1), distinguishing between short-term memory and encoding, storage, and retrieval.

As a starting assumption, B. A. Wilson (1987) postulated that the core deficit of memory impairment is the consolidation, i.e., the storage of

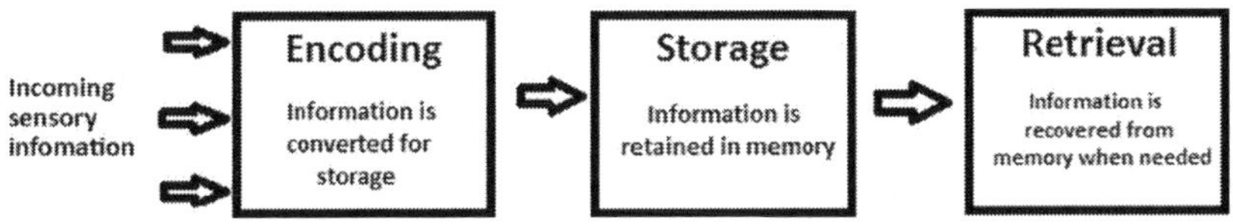

Fig. II.2.1 The Atkinson and Shiffrin (1968) model for memory.

information, and that consolidation deficits cannot be remediated. At the same time, this would mean that encoding is more or less unimpaired. Therefore, it seems reasonable to assume that treatment should be focused on the encoding process. In healthy subjects, using mnemonic strategies seems to be helpful in improving memory performance. Based on results gathered from healthy subjects, the early work of Wilson was focused on the possibility of an internal compensation of the disturbed consolidation process by using mnemonic strategies. In a first experiment, B. A. Wilson (1987) investigated if the PQRST (preview, question, repeat, state, test) method would help memory-impaired patients to retain more information after reading a text. Indeed, she found that this strategy helped the patients retain more information. This study was replicated many years later by Bußmann-Mork, Hildebrandt, Gießelmann, and Sachsenheimer (2000), who also found that the PQRST method was helpful in text reproduction, but did not show generalization to other memory tasks. More recently, Ciaramelli, Neri, Marini, and Braghittoni (2015) documented by experimental variation that patients with mild memory problems due to frontal lesions profited from using the PQRST method to remember text paragraphs. Importantly, the questions developed by the patients spontaneously were just as beneficial as questions developed by the experimenter. This finding supports the argument that the PQRST method can be used by patients independently. However, both replication studies were experimental studies with single interventions and were not designed to show long-term treatment effects.

In the second study, B. A. Wilson (1987) compared four mnemonic strategies in brain-damaged and healthy subjects. The design of the study was again experimental, i.e., it was not a study with a series of treatment sessions. The mnemonic methods she tested included first letter cueing, method of loci, visual imagery, and story building. All subjects started with a "no strategy" condition and the sequence for the different mnemonic strategies was randomized across the participants. In the brain-damaged patients, all four methods improved their memory performance compared to the "no strategy" condition. Whereas visual imagery, method of loci (mentally putting items to be remembered at a specific position), and first letter cueing did not differ in efficacy from each other in this group, the method of story building showed significantly higher effects than any of the other mnemonic strategies in the delayed recall condition (but not in the immediate recall condition, see Fig. II.2.2). B. A. Wilson (1987) interpreted the superior efficacy of the story-building method by arguing that this method combines both visual and verbal aspects. As patients receive more cues through this combination, it might explain the larger gains in memory performance.

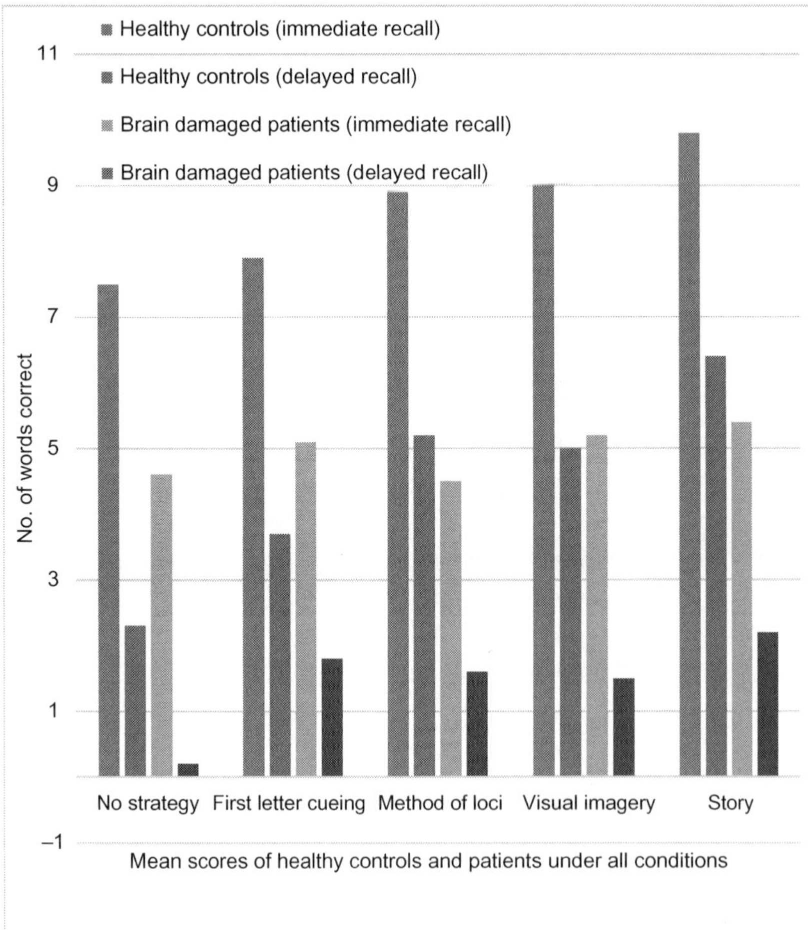

Fig. II.2.2 Impact of different mnemonic strategies on patients with memory disorders (B. A. Wilson, 1987).

On the other hand, immediate recall performance heavily relies on working memory. Working memory was intact in the investigated individuals and therefore the mnemonic strategies were not effective in improving immediate recall (because no relevant improvement was possible).

In yet another experiment, B. A. Wilson (1987) used visual imagery as a mnemonic strategy in the context of a group study. In this study, she distinguished three conditions: learning information without a strategy, learning it with externally cued visual imagery, and learning it with self-generated visual images. The result of this experiment was clear-cut. All groups improved when using visual imagery. The group with severe memory impairment improved only if the visual images were externally cued by the therapist. The moderately impaired group showed some improvement with self-generated visualizations but displayed better improvement for externally cued images. Finally, the group with mild memory impairment profited similarly from both visual imagery conditions.

Note that all these studies by B. A. Wilson (1987), Bußmann-Mork et al. (2000), and Ciaramelli et al. (2015) were experiments that just used the different conditions once in all patients, and they are not really treatment studies with several treatment sessions, randomization, definition of endpoints, etc. Therefore, we do not know if the more-impaired patient groups would have profited from the self-generated visual imagery condition in the case of several treatment sessions focusing on generating such images. Nevertheless, it was a notable experimental finding that moderately to mildly impaired patients could rely on self-organized strategy use and were able to improve their memory performance by visual imagery. This finding was the starting point of a later randomized controlled trial (RCT) exploring the effectiveness of this method in patients with primarily traumatic brain damage.

2.2 An RCT on Visual Imagery to Improve Memory Performance

Kaschel et al. (2002) conducted a treatment study on visual imagery-based mnemonics. Based on the findings of B. A. Wilson (1987), the study focused on mildly memory-impaired patients—assuming that severely memory-impaired patients would not be able to learn and use the mnemonics independently. The study was a multicenter study and the patients were randomized to one of two groups. Nevertheless, it included only a small number of patients (9 memory-impaired patients in the experimental group and 12 in the active control group). The person recruiting the patients was not able to influence the allocation to one of the groups. As an inclusion criterion, a rather high cut-off value for performance in story recall of the Rivermead behavioral memory test (RBMT) was chosen (<16 items) allowing patients with very-mild memory impairments to take part in the study. The patients were blind for the group status. While the therapists were not blind, both groups received active treatment for memory impairment. Therefore, it can be assumed that the motivation for doing the treatment was similar in the two groups. Patients were recruited in the chronic phase of their disease, i.e., years after suffering a TBI or stroke. They were rather young for a neurological patient sample (around 40 years old). The average profile score of the RBMT was high (above 18), showing that they had only minor memory problems. No definite primary outcome variable was defined (except that the treatment should improve memory performance). Assessment involved the RBMT, questionnaires provided to the patients and relatives, as well as some other neuropsychological tests. Although it was a group study, the study followed a SCED design. Therefore, two baseline measures were taken (with a time interval of about 4 weeks in between the measurements), as well as a post-training measure and a 3-month follow-up.

The treatment involved 30 one-hour sessions. First, the patients were motivated by practical examples, which showed that visual imagery may play a role in remembering. Then they were taught to rapidly generate visual images for objects. Afterwards, they learned to generate and retrieve visual images for simple actions. The learning process involved first seeing the objects for a decreasing number of seconds and then creating images of the objects after they had disappeared from the screen. The last step was to create images of objects or actions according to spoken words. The patients were also trained to use the methods in natural settings, and they were instructed how to integrate the method into their daily life.

The control group received treatment as usual. This was composed of attention training, the use of calendars, and the education in different memory strategies adapted to the specific content of the information to be learned and remembered.

As a result, Kaschel et al. (2002) found no impact of the treatment on attention performance, the Wechsler memory scale score, the global RBMT score, immediate prospective memory, and the self-ratings of memory performance (Table II.2.1). However, the imagery patient group showed significant increases in immediate and delayed prose recall, delayed prospective memory performance, and the relative ratings of their memory performance. Significant improvements were still present 3 months after the

Table II.2.1 Test scores of imagery vs. pragmatic training at pre-baseline, post-baseline, post-training, and post-follow-up (Kaschel et al., 2002). Note that the imagery group was the verum group

	Pre-baseline M (SD)	Post-baseline M (SD)	Post-therapy M (SD)	Post follow-up M (SD)
Concentration endurance				
Test d2				
Pragmatic	105.0 (19.9)	110.3 (18.5)	114.7 (18.4)	114.7 (18.4)
Imagery	98.9 (9.6)	102.1 (11.8)	106.3 (10.7)	106.3 (10.7)
Wechsler memory scale				
Pragmatic	–	66.4 (7.1)	67.3 (6.9)	–
Imagery	–	68.0 (10.8)	70.3 (14.7)	–
RBMT				
Profile score				
Pragmatic	18.6 (2.8)	19.7 (3.2)	19.2 (3.2)	20.8 (3.6)
Imagery	19.3 (2.7)	21.0 (3.8)	20.1 (4.5)	22.4 (1.0)
Story immediate				
Pragmatic	8.4 (3.7)	9.4 (3.3)	9.1 (2.1)	9.8 (3.1)
Imagery	9.7 (2.2)	11.2 (3.1)	14.1 (4.5)	15.5 (3.5)
Story delayed				
Pragmatic	6.7 (3.6)	8.1 (3.8)	7.6 (2.8)	8.3 (2.7)
Imagery	7.8 (2.2)	10.4 (2.8)	13.3 (4.5)	14.6 (4.0)
Appointments test				
Immediate				
Pragmatic	2.3 (1.2)	2.5 (1.4)	3.2 (1.6)	2.4 (1.6)
Imagery	2.3 (1.3)	3.3 (1.8)	3.7 (1.8)	4.2 (1.7)
Delayed				
Pragmatic	1.0 (1.2)	1.4 (1.5)	1.6 (1.6)	1.3 (1.2)
Imagery	1.0 (1.3)	1.8 (1.5)	3.1 (1.8)	3.3 (2.2)
Metamemory questionnaire				
Self-rating				
Pragmatic	85.6 (10.6)	87.0 (11.7)	89.8 (10.3)	90.9 (11.6)
Imagery	74.9 (12.7)	75 (14.8)	82.7 (13.8)	88.2 (7.2)

Continued

Table II.2.1 Test scores of imagery vs. pragmatic training at pre-baseline, post-baseline, post-training, and post-follow-up (Kaschel et al., 2002). Note that the imagery group was the verum group—cont'd

	Pre-baseline M (SD)	Post-baseline M (SD)	Post-therapy M (SD)	Post follow-up M (SD)
Relatives' rating				
Pragmatic	74.8 (15.8)	76.7 (18.3)	77.0 (16.2)	78.1 (22.2)
Imagery	73.3 (18.6)	79.1 (21.5)	81.4 (19.4)	91.1 (13.7)

RBMT, Rivermead behavioral memory test.

end of the training. The effect size of the intervention was rather high (in some measures even above 2), although the control group received an active treatment for their cognitive impairment. Apart from some problems in blinding and maybe also in foreseeing the allocation rule, there seems to be no other selection bias (e.g., in reporting or a systematic drop-out of patients from one group). Unfortunately, Kaschel et al. (2002) did not report how many patients from the two groups were able to return to work.

At least two aspects of the study are important and should be mentioned: First, the study showed that patients with mild memory impairments may learn mnemonics such as generating and using visual images without the help of the therapist. This finding replicates the study of B. A. Wilson (1987), who also showed that mildly impaired patients were able to generate beneficial images. Second, the patients of Kaschel et al. (2002) also learned to transfer the method into their daily life (as rated in the questionnaire by the relatives).

From a clinical standpoint, three additional features of the study are noteworthy. First, the high number of treatment sessions (30 h), which is difficult to attain with usual neurological patients in a German rehabilitation unit setting. The second important feature is the young age of the patients (as the incidence of TBI in young people has dropped significantly since then, due to improved safety technologies in cars). Third, they performed high in the RBMT, which shows that they were only very mildly impaired.

2.3 Two RCTs Combining Visual Imagery and Story Generation to Improve Memory Performance

Chiaravalloti, Moore, Nikelshpur, and DeLuca (2013) and Chiaravalloti, Sandry, Moore, and DeLuca (2016) combined visual imagery and integration of the to-be-remembered information in a story into one highly manualized memory treatment program (the "modified Story Memory Technique", mSMT). The whole program consisted of 10 one-on-one treatment sessions, lasting 45–60 min each. During the first four sessions, patients were taught to utilize imagery to facilitate learning. The patients read stories for which they were instructed to remember the target words. They were taught to use mental imagery to increase their memory for this information. The next four sessions were spent using semantic associations combined with mental imagery to improve encoding. The patients read word lists and were taught to first create a highly visualizable story from the list of words. They then were asked to visualize that story according to the mental imagery method they had learned during the first four sessions. The last two sessions focused on generalization to everyday life, and the patients were taught to apply the strategy of the mSMT to real-world memory-demanding tasks.

Chiaravalloti et al. (2013, 2016) tested the effectiveness of this program in two RCTs for patients with TBI (treatment $n = 35$; placebo $n = 34$) and with multiple sclerosis ($n = 86$). The mean age of the TBI patients was 39 years while that of the MS patients was 49 years. The power of the studies was calculated for a $p = 0.05$ and a power of 80%. The studies were tripled blinded, there was an adequate concealment and a follow-up. The control group was treated with similar tasks as the intervention group but without any education of mnemonic strategies. For both studies, the impact of the treatment was also rated from the report of significant others using the Frontal System Behavior Scale (FrSBS). Primary outcome measures were the learning slope of the CVLT and story recall; various neuropsychological tests, and the FrSBS served as secondary outcome measures. The drop-out rate was similar in both groups. As a result, both treatment groups improved signifcantly higher in the learning slope of the CVLT (Fig. II.2.3) and in story recall (Fig. II.2.4). For the TBI patients this latter difference compared to the control group was $p = 0.06$. There were some effects on the FrSBS, but only after analyzing subscores and comparing the two studies on different items (inhibition in the TBI group and apathy in the MS group). These additional analyses for the items of the FrSBS were not corrected for multiple testing.

As already mentioned, Chiaravalloti et al. (2013, 2016) also included a follow-up that took place 6 months after the end of the treatment. A subset of the participants in the treatment group received monthly booster sessions to stabilize the treatment effect over time. Nevertheless, the

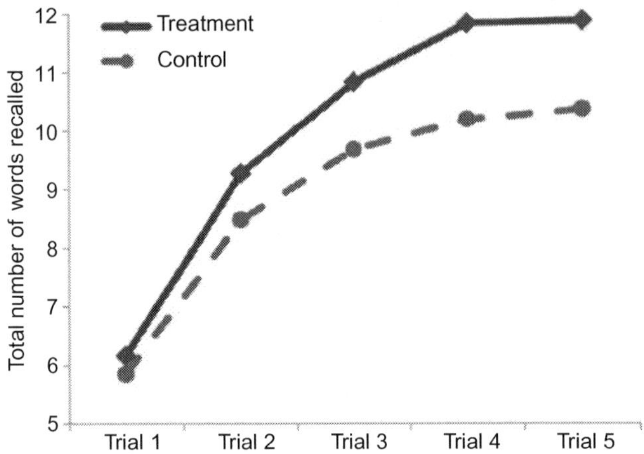

Fig. II.2.3 Improvement of MS patients in CVLT learning after 10 sessions of the modified Story Memory Technique treatment (Chiaravalloti et al., 2013).

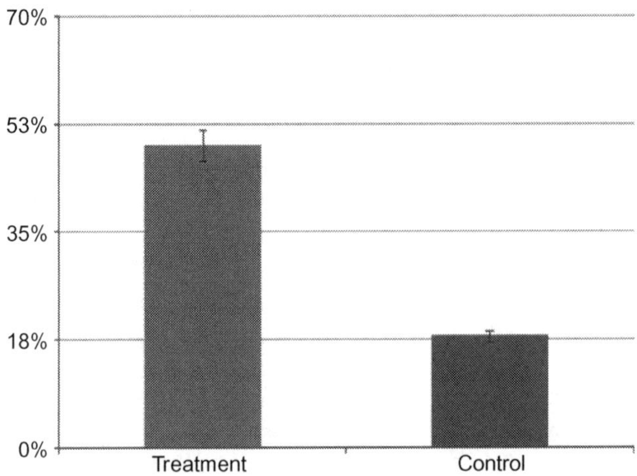

Fig. II.2.4 Improvement of TBI patients in text reproduction after 10 sessions of the modified Story Memory Technique treatment (Chiaravalloti et al., 2016).

performance of both groups was similar at the follow-up assessment. Thus, the benefit of the treatment was present only immediately after the treatment.

Comparing these two studies with that of Kaschel et al. (2002), the number of intervention sessions was much lower (one-third), and the methodological design was almost perfect concerning evidence-based medicine (EBM) standards. The only problem which might be raised is that Chiaravalloti et al. (2013, 2016) did not report all measures that are of interest (e.g., the delayed recall scores of the CVLT). A common feature of all three studies is that the patients were only mildly impaired. Patients in the placebo group in the MS treatment study reached on average >10 words already by the fourth learning trial of the CVLT. From a conceptual point of view, it should be noted that all three studies tried exclusively to improve the process of encoding or acquainting information, and they ignored the processes of retrieval or consolidation.

Nevertheless, according to the EBM standards, the three studies together showed a high level of evidence that in patients with mild memory impairment an internal compensation method via a mnemonic strategy is effective (as also noted in the German guidelines for memory rehabilitation).

3 MEMORY REHABILITATION AS IMPROVING THE ENCODING PROCESS THROUGH MULTIPLE, ADAPTIVE MNEMONIC STRATEGIES

The studies of B. A. Wilson (1987), Kaschel et al. (2002), and Chiaravalloti et al. (2013, 2016) tried to improve memory impairment by teaching patients one or two specific mnemonic strategies. In 1991, Berg et al. published a group study on memory rehabilitation that focused on the optimization of the encoding process through simple strategies and a higher awareness and acceptance of the impairment. The study involved an extended treatment phase, two control groups, and a type of randomization. As in the study of Kaschel et al. (2002), patients of Berg, Koning-Haanstra, and Deelman (1991) were young patients (average age of about 35 years) in a chronic state, suffering mainly from a TBI. The intervention consisted of two sets of 3 weeks of training. Four months after the last training session there was a follow-up assessment (Fig. II.3.1). The study involved three groups: a verum group with 17 patients, an active control group with drill and practice training (11 patients), and a nontreatment group with 11 patients. Berg et al. (1991) stated that the patients were randomly assigned to the treatment and the control condition which is a slightly suspicious statement, as the distribution over the three groups was 17–11–11.

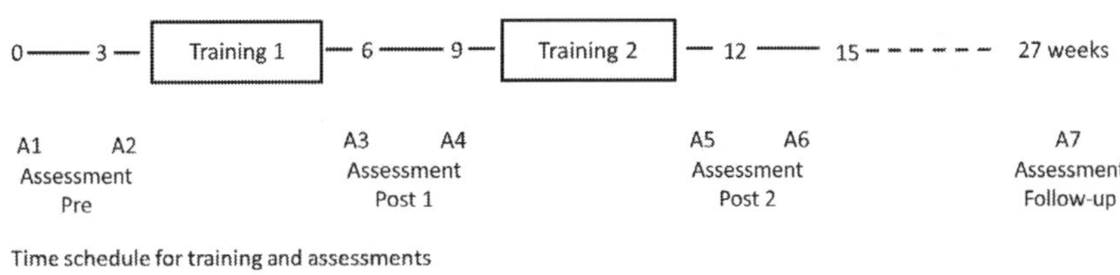

Fig. II.3.1 Design of the Berg et al. (1991) study.

The groups did not differ in respect to age, gender, or length of post-traumatic amnesia. The starting assumptions of the study were quite similar to that of B. A. Wilson (1987). Berg et al. (1991) assumed that as a memory deficiency implies the existence of an impaired consolidation process, it cannot be cured. The treatment therefore focused on the encoding process and the patients were taught to use external memory aids, to pay more attention, to repeat presented information, to make associations between them, and to link input and retrieval situations. At the same time, they were educated to accept their impairment. These rules were illustrated in a textbook and were explained to the participants repeatedly. The specific content of the training was defined by participants, based on the amount of distress it caused them in daily life. During the entire treatment periods, the participants were encouraged to use the mnemonic rules for themselves at home.

The active control group played various memory tasks and games, which were not described in more detail. The primary outcome parameter was a compound score of a 15-word learning test (like the auditory verbal learning test), face name learning, and a recall of a shopping list. The secondary outcome parameters were two attention tasks to control for unspecific improvements during the rehabilitative treatment. The patients of the two intervention groups were also asked if they were satisfied with the treatment using a questionnaire, which contained items to assess mood and quality of life.

Starting with the subjective results of the study, both active groups did not differ: They believed that they had improved in their memory performance that this improvement was applicable in daily life and they felt emotionally better than before the training (Fig. II.3.2). None of the three groups showed any improvement in their attention performance. However, there was a clear improvement in memory performance for the strategy training group compared to the two control groups. The improvement concerned acquisition and delayed recall alike (Fig. II.3.3), it started already during the first training period, continued with almost the same amount of benefit in the second training period, and was still present at the follow-up. Regression analyses showed that the baseline performance and the group membership both contributed to the explanation of the acquisition and free recall performance. Thus, the study showed that 18 h of memory training significantly improved performance compared to an active control group and an untreated group. On the other hand, the study also showed that the treatment approach had *no* differential impact on mood, quality of life, and experienced increases in memory performance, because both active groups

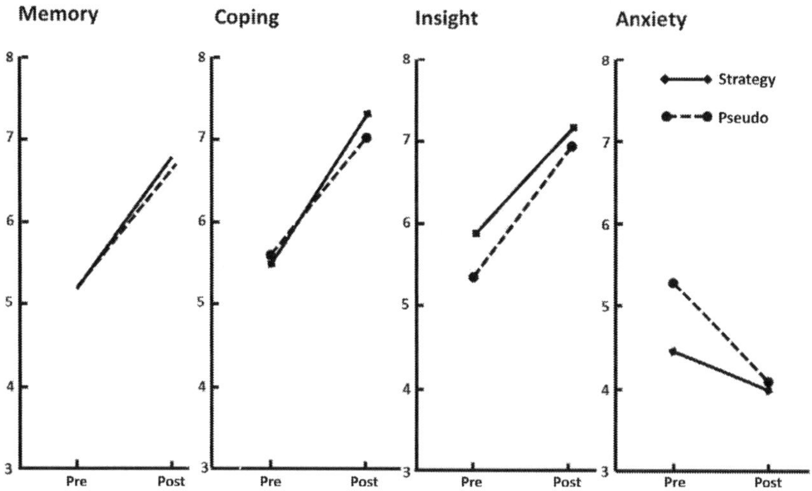

Fig. II.3.2 Subjective improvement in the strategy and the drill and practice group (Berg et al., 1991).

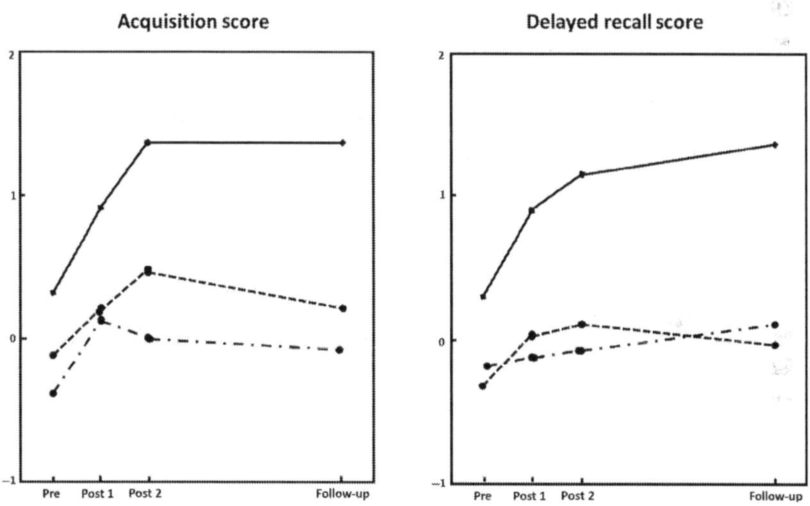

Fig. II.3.3 Improvement in memory functions in the Berg et al. (1991) study.

reported a similar improvement. On this basis, self-reports can only play a minor role in evaluating neuropsychological treatment success.

Let us try to apply the different criteria of EBM to this first and very impressive neuropsychological memory rehabilitation study. The first question is: Did the study show an improvement in experimental parameters, in clinical aspects or in health economical aspects? The answer to this question is not as clear-cut as one might think, because of the problem of how to define a clinically relevant improvement in memory performance. After all, Berg et al. (1991) used memory tests for their study and one might argue that a relatives' rating scale for daily performance is not a sufficient measure because of being aware of the provided treatment. The second question concerns the level of evidence which is reached by this study. According to the method section, Berg et al. (1991) randomized the patients to one of the three groups. On the other hand, no information is given on how the randomization worked and why the group sizes differed systematically. Moreover, there is no information about concealment, about drop-outs and about blinding of the assessing neuropsychologist. Because the two active groups differed systematically (the placebo treatment was conceptualized as such a treatment), no blinding of the therapists was possible. On the other hand, because the patients estimated the effectiveness of both interventions similarly, it appears that the patients were blind to their group status. Thus, the study was at least single-blinded. Following the Cochrane collaboration group, the next question is if the study shows an attrition bias (loss of specific outcome data). This is obviously not the case. On the other hand, there might have been a selective reporting bias, because memory questionnaires were passed to the patients and to a close relative, but the latter results were not reported in the paper.

Summing up these methodological problems, it seems fair to argue that the study's EBM level is Level IIa and not Ib. The main reasons for this classification are the lack of double-blinding and the systematic difference between the group sizes. This difference argues against a true randomization or argues for a systematic loss of patients in the two control groups. These potential factors might have influenced the significant difference of memory performance between the groups. Nevertheless, the study of Berg et al. (1991) was clearly a high-standing study on memory rehabilitation, considering that the concept of EBM was not yet developed at the time.

As argued above, the methodological framework is only one important aspect that decides whether a treatment study is a high-level study. The other aspect concerns the conceptual setup of the study. In this respect, let us consider the strategy taught to the patients to improve encoding, and how the improvement in memory performance was assessed. Assessment was undertaken using word list learning and face name association learning. Both tasks do not allow the use of many of the strategies taught to the patients. For example, at least during standard oral presentation of the word list it is not possible to take more time for learning and repeating information is also not possible or will have detrimental effects on the performance. External memory aids were forbidden during the assessment. Therefore, while the Berg et al. (1991) study showed empirically that improvement of memory is possible, on a conceptual level it remains unclear how this improvement was achieved. For clinical purposes, it should also be mentioned that the patients were young, that their impairment was most often due to TBI, and that the patients received 18 total face to face individualized sessions of 60 min each (which is a greater amount as usual during standard clinical practice). A replication study by Doornhein and De Haan (1998) included 12 stroke patients that were randomized into two groups. This study showed an improvement in specifically trained memory processes, but no transfer to untrained memory functions. The group consisted of stroke patients who were about 16 years older than the TBI patients of Berg et al. (1991), which may explain the difference between the studies. Furthermore, with case numbers of 6 versus 6, the impact of the intervention must have been expected to be very strong to yield any significant results at all.

4 DEEP ENCODING AND TRANSFER APPROPRIATE PROCESSING AND MEMORY REHABILITATION

4.1 Deep Versus Shallow Encoding

B. A. Wilson (1987) and Berg et al. (1991) both started with the assumptions that consolidation of information cannot be improved, that encoding of information is unimpaired, and that mnemonic strategies can therefore be used to help memory-impaired patients cope with their problems. B. A. Wilson (1987) was skeptical as to whether mnemonic strategies will generalize to daily life situations, whereas the study by Berg et al. (1991) showed that the patients believe that memory training is helpful in such situations (although the active control group believed that too). In the study by B. A. Wilson (1987) the one strategy that was particularly successful was the story method. For the Berg et al. (1991) study we can assume some strategies to be unimportant, because the assessment strategy excluded their usage. These include paying more attention (attention performance did not change during the intervention), taking more time and repeating information. Moreover, these three strategies are unspecific for encoding and retrieval of information, because any performance (not specifically memory performance) will be positively influenced by them. Excluding these strategies as relevant for the explanation of the gains in memory performance it might be that focusing on semantic associations is the common cognitive mechanism that explains the success in the B. A. Wilson (1987) and Berg et al. (1991) studies. However, there is still some contradiction on a conceptual level here: Both studies did not propose to focus on *endogenous and more or less automatic* encoding processes for improving memory performance but instead designed a treatment that *added* some mnemonic components to the encoding process. At the same time, the findings do not really fit this conceptual assumption, because the strategy to evaluate the treatment success did not allow for any additional strategic cognitive process. From a cognitive neuropsychological standpoint, one is inclined to argue that the success of the studies relied on changing an automatic cognitive process, although the intention was to teach the patients a conscious cognitive strategy.

Is it possible to identify such inherent and automatic subcomponents of encoding? As mentioned in the paragraph above, the study by B. A. Wilson (1987), which compared four different mnemonic strategies, pointed in the direction that semantic associations (the core element of the story method) might be such a subcomponent and could play an important role in memory performance. Previous work with amnesic patients, mainly suffering from Korsakoff's syndrome, revealed a similar result. Already in 1979, McDowall (1979) published an important paper that was based on the earlier work of Cermak & Butters (1972, 1973, 1974), and analyzed the effect of encoding and retrieval conditions on the memory performance in Korsakoff's patients, a group of patients suffering from severe memory deficits due to malnutrition (lack of vitamin B1 and B12) often as a consequence of alcohol abuse. In a first experiment, McDowall (1979) compared three word list learning conditions: In the semantic condition (S) the words were presented together with category labels and the patients had to decide if the category label was the correct one. In a second condition (NS) patients were instructed to decide whether there was the letter "e" in the presented word. In the third, neutral, condition (N) patients were just instructed to remember the words. During the retrieval condition, there were no cues. Fig. II.4.1A shows the results.

As can be seen in the figure, Korsakoff's amnesics improved significantly in their memory function when they were cued to focus on the meaning of the word (although they did not reach the level of performance of alcoholics without memory deficits), but without such instruction (condition N) they were unable to use the semantic information inherent in the studied words for increasing memory performance. The question then is if a lack of spontaneous semantic processing (deep encoding in contrast to shallow encoding, where attention is directed towards the characters of the word)

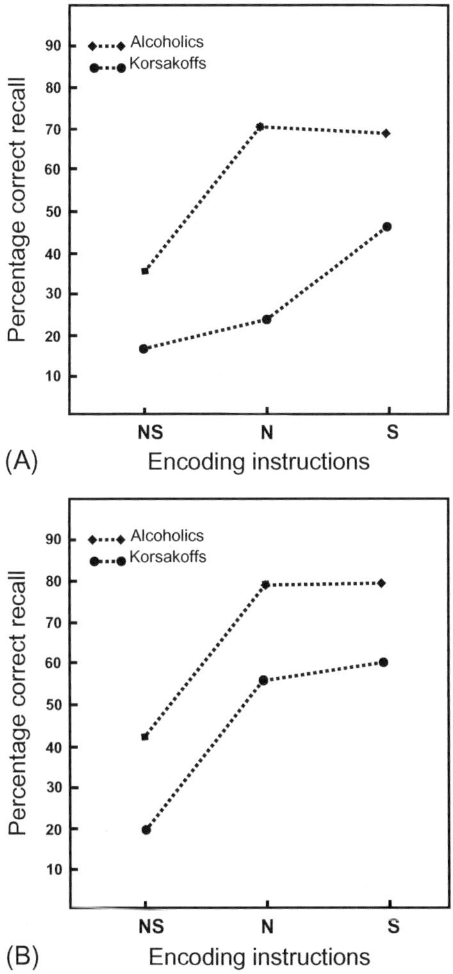

Fig. II.4.1 Results of the McDowall (1979) investigation on memory performance of patients with Korsakoff's amnesia. (A) Mean percentage correct recall for Korsakoff and alcoholic groups as a function of encoding instructions; (B) Mean percentage correct recall for Korsakoff and alcoholic groups in the presence of category cues at recall, as a function of encoding instructions. (NS = nonsemantic instructions, N = no instructions, S = semantic instructions).

during encoding may be one cause of the memory deficits in Korsakoff's amnesics. McDowall (1979) performed a second experiment where he varied the retrieval process. The three encoding conditions were the same as in the first experiment, but during retrieval the patients received semantic cues (like "there were fruits" and "there were animals"). Fig. II.4.1B (right side) shows the results of the second experiment. Korsakoff's amnesics encode the semantic meaning of words, irrespective of being cued to do so, but they do not spontaneously use such cues for retrieval. Therefore, providing them with such cues during the retrieval process improved their memory functions (as compared to the no-cue retrieval condition during experiment one) to the same level as providing them with such cues during encoding. On the other hand, cueing during encoding *and* during retrieval did not exert an add-on effect, and the performance of the patients was like that in the neutral condition. Thus, semantic cueing during retrieval or during encoding is effective enough to increase memory performance and semantic, deep encoding is not impaired (at least not implicitly). The impairment concerns the active generation and use of information that is implicitly represented but is not being used.

Cermak et al. (1974), using a different experimental design, showed that it depends on the quality of contextual shifts how severe Korsakoff's amnesics are impaired in responding spontaneously and automatically to semantic meaning. They used the method of proactive interference, which means that the participants had to learn four different lists of the same category (digits or words stemming from the same category; e.g., animals). Presenting the fifth list implied a change of the category (i.e., if the participants learned digits previously, they now had to learn characters or if they had to learn animals before, they now had to learn examples from the category of furniture). Fig. II.4.2 shows the results of the learning experiments for alcoholics (top line in each the figure), who served as a control group, and Korsakoff's amnesics (bottom line in each figure).

The results document that learning items from a common category in different lists leads to a decrease in memory performance—the proactive interference effect. A comparison of Figs. II.4.2B and II.4.2A, and of Figs. II.4.2C and II.4.2D shows that Korsakoff's amnesics highly profited from the nonsemantic shift from digits to characters (the "release of proactive interference"). At the same time, they were not influenced by a semantic shift between word categories at all. On the other hand, nonamnesic alcoholics showed a release effect also for a shift between semantic categories. Subsequent studies found that increase of proactive interference and reduced release of proactive interference might relate to frontal lesions (Griffiths et al., 2005; Sagar, Sullivan, Cooper, & Jordan, 1991; M. L. Smith, Leonard, Crane, & Milner, 1995). Moreover, the effects are especially strong if recall is effortful, and the lack of any automatic release of proactive interference (as in the Cermak et al., 1974 study) is an exception.

In some respects, the findings of McDowall (1979) are similar to the results of B. A. Wilson (1987) investigating the effects of visual imagery on the

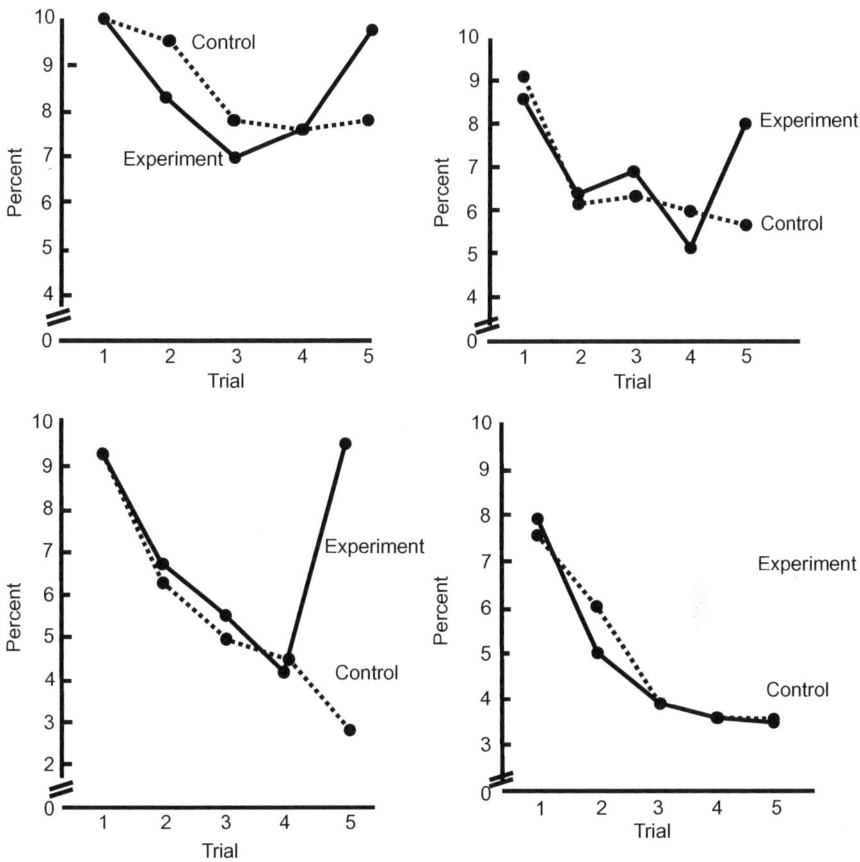

Fig. II.4.2 The impact of proactive interference and release of proactive interference on list learning in Korsakoff's amnesics (Cermak et al., 1974). (A) Probability of recall following an alphanumeric shift for alcoholic controls. (B) Probability of recall following an alphanumeric shift for Korsakoff patients. (C) Probability of recall following a taxonomic shift for alcoholic controls. (D) Probability of recall following a taxonomonic shift for Korsakoff patients.

memory performance of amnesic patients. In both investigations, the patients profited from externally provided cues, but failed to develop such cues spontaneously. One major difference between the investigations is that the results from McDowall (1979) argue that cues may act during retrieval as well as during encoding, whereas B. A. Wilson (1987) focused exclusively on encoding. If it is possible to recover the semantic structure of information during retrieval, it might be possible to teach patients a more efficient strategic search. This in turn might increase their memory performance. Even more important, semantic processing is an endogenous part of encoding and retrieval processes and it may be deficient on different levels (on that of representing or accessing the information). The findings of McDowall (1979) argue for an access deficit—a lack of strategic use of acquired information. The results of Cermak, Butters, and Gerrein (1973) and Cermak et al. (1974) argue for a representation deficit, because proactive interference is an automatic process inhibiting memory performance. The lack of release from proactive interference in the Korsakoff's amnesics seem to indicate that they do not represent the semantic information inherent in the list shift, but that they do show a release effect if the difference between the lists do not rely on semantics.

Together these findings argue that semantic processing is an automatic aspect of encoding and that memory performance is influenced by it. Therefore, it might be not necessary to teach patients a completely new and additional strategy to improve their memory. Rather, patients just need to focus on the semantic associations between the information to be remembered. The story method (relying on semantic structure) and one of the strategies of Berg et al. (1991) (i.e., to make semantic associations) underline this possibility, because both show the efficacy of semantic structuring in improving memory performance.

Before reviewing studies on the efficacy of semantic structuring ("deep encoding") in the improvement of memory performance (see Fig. II.4.3 for different aspects of levels of encoding), it should be noted that the original explanation of the memory deficits in Korsakoff's amnesia, which was based on the relation between semantic encoding and memory performance has been criticized during the 1990s (Morris, Bransford, & Franks, 1977; for a review see Ekuni, Vaz, & Bueno, 2011).

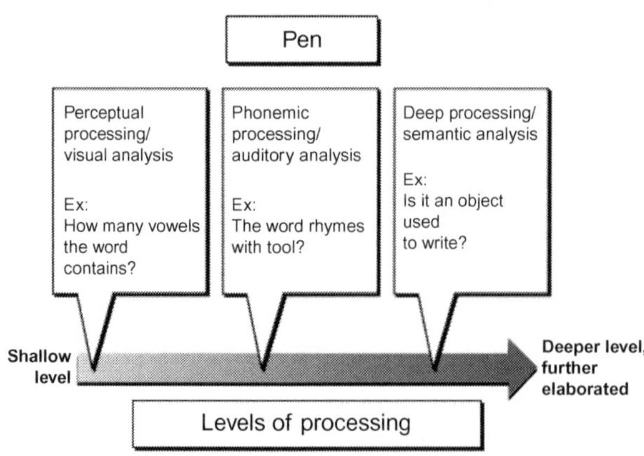

Fig. II.4.3 The levels of processing model for encoding (Ekuni et al., 2011).

Two aspects of the arguments raised against this theory should be briefly discussed, because they are relevant for memory rehabilitation. The first argument concerns a strong variant of the deep encoding explanation for the memory deficits found in Korsakoff's amnesics. The strong variant argues that the impairment is *caused* (exclusively) by a lack of deep encoding. As the Korsakoff's amnesics do not process the information (spontaneously and/or automatically) on a deep level (neither during encoding nor during retrieval), they show a memory impairment. This assumption has proven wrong as can be seen by the two figures from McDowall (1979): The difference between the alcoholics and the Korsakoff's amnesics remains the same after semantic cueing. It is just the case that the difference even increases *without* semantic cues during the encoding or the retrieval process. Moreover, the second experiment of McDowall (1979) showed that some semantic structure of the information seems to be unconsciously present even in Korsakoff's amnesics. Their main problem is in using self-generated search schemata to recover the semantic structure of the information, because providing them with such cues led to an improvement of memory performance.

In interpreting the results of the McDowall (1979) study one should be aware that the effects may also be influenced by the difficulty of the task. The figures suggest that semantic encoding processes work as efficiently on an implicit level in alcoholics as in Korsakoff's amnesics. However, the study by Cermak et al. (1974) on proactive interference, i.e., with a more difficult task, shows that semantic processing actually seems to be disturbed to a certain degree in Korsakoff's patients. Nevertheless, the strong variant of the deep encoding explanation for memory impairment overstretches the interpretation of the findings.

The second argument against the deep encoding theory unfetters it from the limitations of semantic processing. Any processing appropriate for the retrieval process leads to a higher memory performance. If the task is to remember meaningful words, deep encoding will improve memory performance because retrieval concerns semantics. If the task is to remember words with specific characters (or specific "shallow features"), a shallow processing (looking for characters of words and not for their meaning) will improve memory performance. This has led to the formulation of the *transfer appropriate processing* model. According to this model, memory performance is optimal in the case that the context of encoding and retrieval is maximally similar, and deep encoding is only appropriate if meaningful representations must be remembered (which is most often the case).

The transfer appropriate processing model is also endorsed by other memory studies. Context-dependent memory is a term that means that memory functions perform best if the context of the retrieval situation (e.g., stress, being on drugs, etc.) resembles the context of the encoding situation. Subsequent studies showed that it is enough to simply create the same internal state during retrieval that was present at the time of encoding. The main assumption of the transfer appropriate processing principle is that retrieval of an episodic memory involves the reinstatement of *cognitive processes* active at the time of encoding and storage of the representation (Rugg, Johnson, & Uncapher, 2015). Therefore, using the same mental set and strategy for retrieval that is used for encoding will help in recalling more information.

It should be clear that apart from the similarity of the encoding and retrieval processes, remembering involves more than just identical neuronal activation. Otherwise, it would be impossible to distinguish between experiencing, imagining, and remembering. Remembering involves the activation of control processes or a retrieval mode and an active reconstruction of the event in question. We will discuss such processes like "post-retrieval monitoring" or "distinctive heuristic" in the section of retrieval-based approaches to memory rehabilitation. When reviewing the results of treatments that rely on the transfer appropriate processing principle (see below), one should keep in mind that these additional processes may also play a significant role.

Summarizing the results on deep encoding in Korsakoff's amnesics, two findings are relevant for memory rehabilitation: First, semantic encoding plays a major role in memory performance, at least if the content (and not the visual aspects) of words is what is to be remembered. Semantic encoding seems to be an automatic aspect of memory, as shown by the proactive interference experiments, and therefore it is not something that is added to memory processes. In other words, semantic encoding is inherent to memory and is not an additional strategy (a mnemonic). Second, there is another automatic process that helps in improving memory function: using the *same* strategy during retrieving information as during encoding information. Again, such transfer appropriate processing is inherent to memory and does not need to be learned and used additionally as in the traditional mnemonic strategies.

Then going back to the Atkinson and Shiffrin (1968) model for memory, we can add some additional components which might help to rehabilitate memory-impaired patients (Fig. II.4.4).

Fig. II.4.4 The Atkinson and Shiffrin (1968) model for memory augmented by semantic structuring and transfer appropriate processing.

4.2 Treatment of Mild to Moderate Memory Disorders by Improving Semantic Structuring and Transfer Appropriate Processing

Is it possible to train semantic structuring and transfer appropriate processing in mildly to moderately memory-impaired patients? If the answer is yes, what may be the constraints? During the first decade of this century, we published a series of treatment studies using these principles. The first study (Hildebrandt, Bussmann-Mork, & Schwendemann, 2006) compared three different patient groups (Table II.4.1). One group was trained for semantic structuring and transfer appropriate processing (POT), the second group learned memory strategies (ST), and the third group (controls) received low dose memory treatment "as usual" (which in some respect resembled the training of the first group). The first group was informed that they should not utilize the sequence in which the information was presented in their attempt at recall—a spontaneous strategy of many memory-impaired patients to cope with their problem. Then they were instructed to try to structure the information into semantic groups. They were told to stick to this semantic structure during encoding and retrieval, irrespective of their subjective beliefs on which strategy was optimal for them, and irrespective of the first information they remembered during retrieval. Instead of focusing on the first piece of information that came to mind, they were told to recover the headings used to group the information during encoding and to search for as many correct items for a heading as possible before shifting to the next category. The patients practiced these principles in a group and received "homework" to practice it several times until the next session took place (which was usually the next day, because the study was conducted in a rehabilitation unit). During the group sessions, the therapist took care that the patients followed the principle to develop a semantic structure and stuck to it during encoding and retrieval. Moreover, the group had intensive training in detecting semantic structure and in developing semantic associations. The other group received the same number of treatment sessions, but they

Table II.4.1 Group characteristics of the transfer appropriate processing (POT) and semantic structuring (ST) study (Hildebrandt et al., 2006) at baseline testing

	Controls	POT	ST
N	16	24	22
Sex (m/f)	12/4	12/12	15/7
Etiology	Stroke: 13	Stroke: 17	Stroke: 11
	TBI: 1	TBI: 2	TBI: 4
	SAB: 1	SAB: 3	SAB: 2
	Others: 1	Others: 2	Others: 5
Ages (in years)	62.44 (11.85)	63.21 (12.09)	56.45 (11.23)
Time since illness (months)	2.34 (3.37)	4.60 (10.74)	9.74 (30.28)
Duration of rehabilitation (days)	26.94 (8.84)	25.04 (8.41)	25.95 (8.84)
Verbal IQ	99.63 (7.7)	99.8 (10.3)	95.7 (11.0)
Performance IQ	103.9 (10.6)	105.4 (9.2)	103.0 (12.3)
Digit-Symbol-Test	24.4 (9.1)	28.8 (11.6)	27.4 (12.9)
LGT Map learning	11.33 (7.41)	11.55 (4.66)	10.57 (4.61)

Figures in parenthesis = Standard-Deviation from the mean. Figures before parenthesis = mean. Abbreviations: *POT*, process oriented training; *ST*, Strategy training; *m*, male; *f*, female; *TBI*, traumatic brain injury; *SAB*, subarachnoidal bleeding; *IQ*, intelligence quotient; *LGT*, Lern-und Gedächtnistest.

instead focused on various strategies for name-face association learning, the PQRST strategy, etc. One may compare the second group to the intervention group of the Berg et al. (1991) study, who also received education and training in several strategies and adapted to their individual and contextual needs. As already mentioned, the third group received low-dose memory training. Most of the included patients suffered from a past stroke. Additionally, all patients were over 60, which is a typical age for patients in German neurological rehabilitation units. The time since experiencing the stroke or disease varied, but it was generally around 2 months. This means that most of the patients were still in a post-acute phase of their disease.

After the training both intervention groups showed significant improvements in memory functions compared to the control group. Improvement concerned the results of subtests of the CVLT, but it also appeared in text reproduction and in verbal fluency tasks.

Whereas the performance increase in the learning trials of the CVLT was quite similar in both treatment groups, the group that was trained in

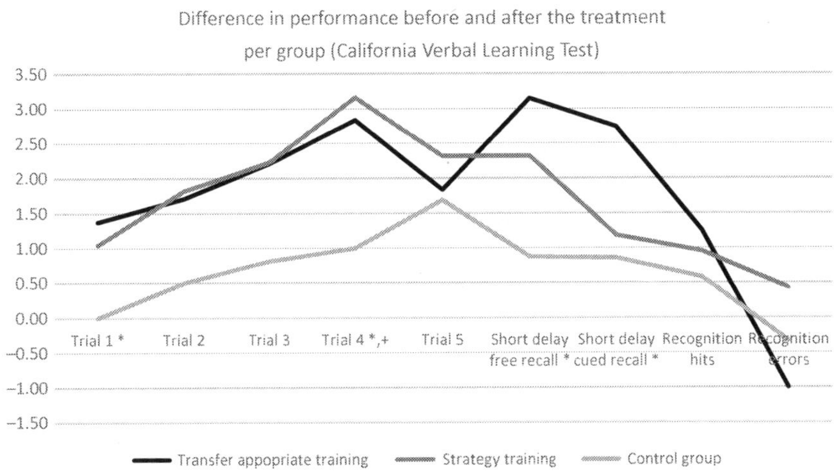

Fig. II.4.5 Improvement during learning and recall in the CVLT for the different groups of Hildebrandt et al. (2006). * $p < 0.05$ in comparison between the transfer appropriate training and the control group. $^+$ $p < 0.05$ in comparison between the strategy training and the control group.

semantic structuring and transfer appropriate processing improved more in delayed recall measures and especially in semantically cued free recall (Fig. II.4.5). The improvement of semantically cued free recall was almost statistically significant ($p < 0.06$). Using the CVLT, that allows an investigation of semantic clustering, the study demonstrated that this group learned to group words according to their semantic structure during delayed recall.

Three additional aspects of this study should be mentioned. The first aspect is that attention testing showed no difference for the three groups at t1 and t2. Therefore, the improvement seemed to be related to memory function and not to a global recovery (many of the patients were in a sub-acute state). The second is that the semantic structuring and transfer appropriate processing group also improved in text reproduction. Strategies in text reproduction were taught to the other intervention group, which trained different strategies, but text reproduction was not part of the treatment of the semantic structuring and transfer appropriate processing group. Thus, one can argue that this training showed a type of transfer effect onto a more clinically relevant performance. Furthermore, dividing the patients into groups of different severity of impairment showed that significantly more patients of the semantic structuring and transfer appropriate processing group achieved a status change from moderately impaired to mildly impaired (Fig. II.4.6).

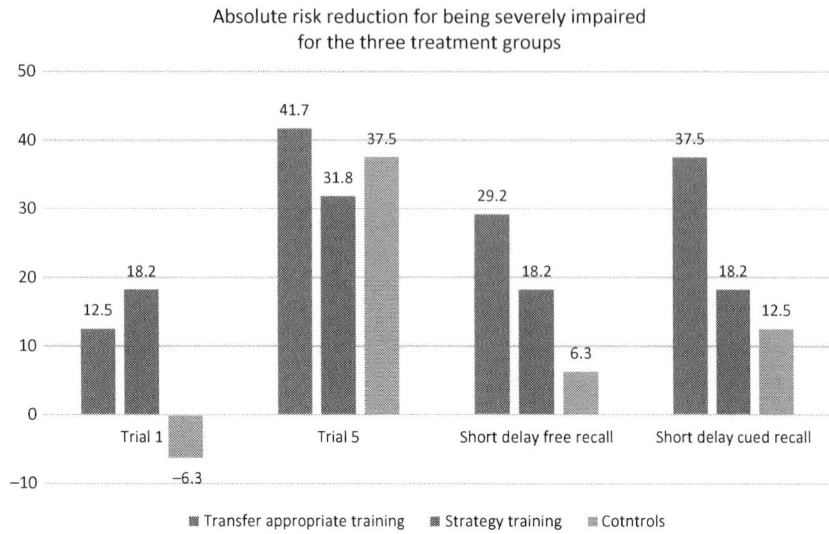

Fig. II.4.6 Percentage of patients who improved from severely impaired to moderately or mildly impaired according to their group status (Hildebrandt et al., 2006).

In terms of EBM, the design of the study of Hildebrandt et al. (2006) can be criticized in a few ways. First, there was not a complete randomization of all patients, as we stopped randomization after having recruited about half of the patients and then focused on increasing the number of the two verum groups and tried to match them for gender, age, and education. Moreover, there was no blinding of the assessor, no systematic presentation of the drop-outs, and also no power calculation. In this respect, only the difference between the two treatment groups can be considered a high-level Level Ib result. The difference of the two treatment groups compared to the control group represents only a Level IIa result. Last, but not least, there was not a follow-up assessment. On the other hand, there was no selection bias in reporting. Furthermore, the effect size of the treatment (already standardized on the improvement of the control group) was >1 for the free recall performance in the CVLT and >0.5 in text reproduction. These are clearly impressive results.

We were able to replicate the findings of Hildebrandt et al. (2006) in the years after the study's publication using different variations of the therapeutic method (Hildebrandt, Clausing, Janssen, and Moedden, 2007; Hildebrandt, Gehrmann, Modden, & Eling, 2011). To mention one other study, Hildebrandt et al., 2011 transferred the principles of the group treatment into a software program. This program was run with single patients in the presence of a therapist, who intervened if the patients started to deviate from

Table II.4.2 Demographical data of the two treatment groups of Study 1 in Hildebrandt et al. (2011)

	Computer treatment		Standard group treatment	
	2 stroke; 5 SAB, 5 hypoxia		8 stroke, 2 ICB, 2 SAB, 1 tumors, 2 hypoxia	
Etiology	**Mean**	**SD**	**Mean**	**SD**
Age	50.8	10.6	57.9	12.0
Sex	M: 9	F: 3	M: 13	F: 2
Time since onset (in month)	1.8	1.2	1.2	0.3
Duration of rehabilitation (in days)	29.3	7.4	24.7	7.7
Premorbid intelligence (IQ)	101.7	11.3	95.1	6.4
Patients below the 10th percentile of the SDFR	≥1: 9; 2: 0; 3: 1; 5: 2; 7: 0		≥1: 7; 2: 5; 3: 1; 5: 1; 7: 1	

Abbreviations: *ICB*, spontaneous intracerebral bleeding; *SAB*, subarachnoidal bleeding; *SDFR*, Short delay free recall of the California Verbal Learning Test.

the treatment principles. In this study, the nurse of the rehabilitation unit judged if the patients improved in their memory performance while being blinded for the treatment conditions. Table II.4.2 shows the number of included patients and Fig. II.4.7 displays the results of the treatment. Again, a significantly higher improvement for the semantic structuring and transfer appropriate processing group could be demonstrated (Fig. II.4.7). The effect size of the treatment for the short free recall was >1, even after subtracting the improvement of the control group from that of the treatment group. The effect size for the improvement of story recall was lower, i.e., about 0.3, and the statistical analysis did not show a significant difference between the groups. One explanation for a lack of generalization to story recall might be that the training intensity was much lower than in the Hildebrandt et al. (2006) study. On the other hand, the nurses judged that the experimental group improved in their daily life memory performance.

A recent study by Miotto et al. (2013) based on a similar treatment intervention used not only behavioral measures, but also functional brain imaging. It showed that patients with memory impairments due to frontal lesions can learn semantic structuring quite easily (at least for a single and cued application), and that the effects of using such a strategy also changed their fMRI

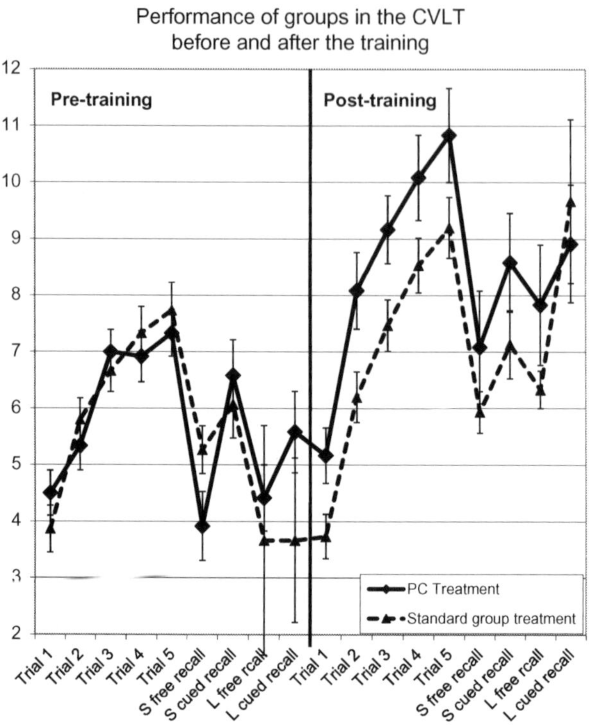

Fig. II.4.7 Improvement in the CVLT (Hildebrandt et al. (2011), first study).

BOLD responses. Twenty-one adult patients with post-acute prefrontal cortex (PFC) lesions, 12 with left dorsolateral PFC (LPFC) lesions, and 9 with bilateral orbitofrontal cortex (BOFC) lesions were included in this study. Scanning was carried out twice, before and after a 30-min semantic organizational strategy training outside the scanner on the same day. Subjects were instructed to look at words presented onto the screen and try to remember as many words as they could. They were assessed before and after the cognitive semantic structuring training with additional neuropsychological tests outside the scanner. The training consisted of teaching participants to apply semantic organizational memory strategies to a set of different word lists. They were instructed to first organize the words into categories and then to memorize and retrieve the words according to each category to improve recall. Recall performance increased after the training of semantic structuring (Table II.4.3) and the fMRI experiment showed increased BOLD responses after the training in the left premotor and prefrontal areas (Fig. II.4.8), i.e., in brain regions responsible for semantic processing and verbal working memory.

Table II.4.3 Improvement of free recall in patients with frontal lesions: Clinical characteristics and performance on experimental memory tasks measured before and after training of patients with BOFC and LPFC lesions (Miotto et al., 2013)

	BOFC	LPFC
Age	45.73 (10.24)	39.57 (10.89)
Years of education	8.91 (2.74)	8.93 (3.52)
Full IQ (WASI)	97.00 (7.55)	99.79 (10.12)
Digit span percentile (WASI)	30.00 (18.48)	40.93 (21.05)
Verbal recognition percentile (WRMT)	36.55 (21.34)	48.43 (27.48)
Visual recognition percentile (WRMT)	41.91 (21.96)	52.07 (27.48)
RAVLT immediate total recall percentile	19.3 (7.1)	9.2 (8.2)
RAVLT delayed recall percentile	18.6 (1.7)	21.8 (2.3)

	BOFC		LPFC	
Experimental memory tasks	**Pre-training**	**Post-training**	**Pre-training**	**Post-training**
Free recall*	31.6 (15.99)	46.6 (11.52)	29.0 (14.07)	46.29 (12.5)
Semantic clustering index*	0.51 (0.14)	0.56 (0.16)	0.25 (0.14)	0.27 (0.14)

Results are expressed as mean (SD).
Abbreviations: *BOFC*, bilateral orbifrontal cortex; *LPFC*, lateral prefrontal cortex; *WASI*, Wechsler Adult Short Intelligence Scale; *RAVLT*, Rey Auditory Verbal Learning Test; *WRMT*, Warrington Recognition Memory Test.
*Significant training effect at $p < 0.001$.

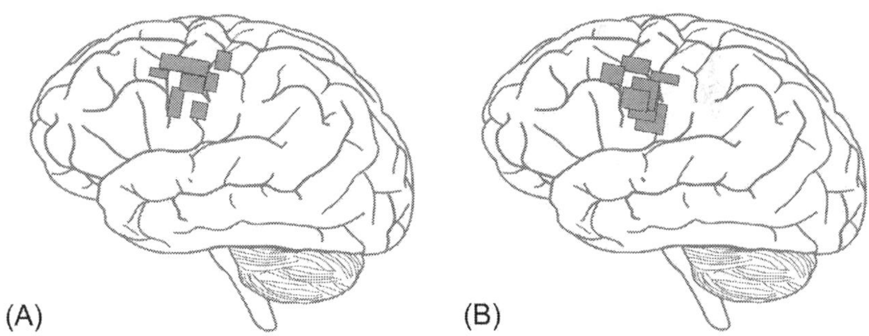

Fig. II.4.8 Brain regions showing increased activity in the posttraining relative to pretraining session in patients with LPFC lesions. (A) In the related-nonstructured words condition, greater activation after training was observed on the left hemisphere in inferior frontal gyrus (−47, 7, 15), precentral gyrus (−51, 0, 20), and insula (−40, 7, 9). (B) In the related-structured words condition, greater activation after training was observed in the left hemisphere in inferior frontal gyrus (−47, 7, 15), precentral gyrus (−36, −7, 26), and anterior cingulate (−18, 14, 36). $P < 0.01$ cluster level (Miotto et al., 2013).

Nyhus and Badre (2015) reviewed fMRI studies on memory retrieval in healthy subjects focusing on the PFC. Retrieval effort was correlated with increased activation in the ventrolateral PFC. While the activation was located somewhat more anterior than in the Miotto et al. (2013) study, this difference may be explained by the fact that the latter study included patients with lesions in the frontal cortex. According to Nyhus and Badre (2015) there is evidence for two functionally specific areas that are both active during any retrieval attempt: the anterior ventrolateral and the mid ventrolateral PFC. The first one is highly involved in selection and combination of relevant features to increase the efficacy of memory search. An example of such a selection process would be the binding of processes and contextual information (which were present during the encoding period) into one search schema and looking for the stored information (that was encoded earlier) which maximally fit this combination. On the other hand, a search for memory contents by anterior ventrolateral prefrontal activation using incomplete information bears the risk that similar but irrelevant memory traces get activated. Therefore, Nyhus and Badre (2015) argue that the mid ventrolateral PFC is involved in post-retrieval monitoring of reactivated memory traces. The increased risk of remembering faulty information may also play a role in memory rehabilitation and will be discussed below. At this stage, it is enough to state that there is converging knowledge that semantic structuring and organization play an important automatic role in encoding and retrieval, and this argues for the possibility of increasing its efficacy through rehabilitative training attempts.

The meaning of semantic structuring for memory performance in amnesic patients was also shown by Sheldon, Macdonald, and Schweizer (2012) (for earlier findings on this topic compare della Rocchetta and Milner (1993)). In this study, the memory performance of patients with nontraumatic aneurysmal subarachnoidal hemorrhage was compared with healthy controls. The patient group was divided into patients with anterior or posterior aneurysms. Patients with SAB suffered from memory problems, with the posterior group suffering even more than the anterior group (Fig. II.4.9). Sheldon et al. (2012) used semantically organized and semantically nonorganized lists for their investigation. They were able to show that semantic organization of the word list led to a significant improvement in list learning, helping the anterior patients to achieve a similar performance as the healthy controls. For the unorganized list, the performance of the anterior group dropped to that of the posterior group, i.e., significantly impaired compared to the control group.

Note that the results of Sheldon et al. (2012) on amnesic patients with anterior subarachnoidal aneurysmal hemorrhage, the results of della

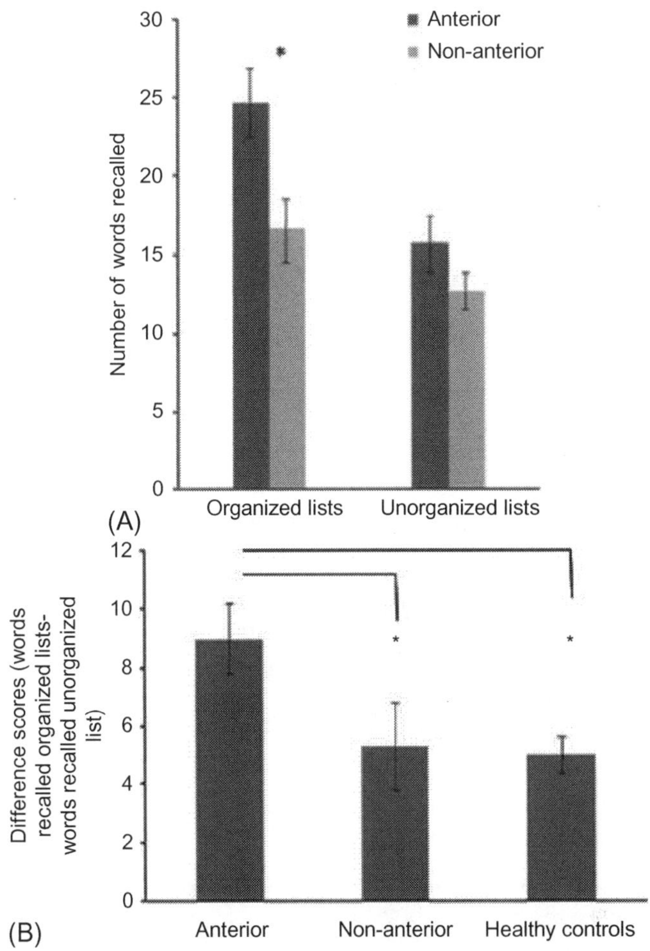

Fig. II.4.9 (A) The effect of semantic list organization on recall performance of patients with anterior and posterior SAB (Sheldon et al., 2012). (B) Difference score for recall of semantically organized or not-organized lists in patients with anterior and posterior SAB or healthy controls (Sheldon et al., 2012)

Rocchetta and Milner (1993) on patients with cortical resection to treat epilepsies, and finally those of McDowall (1979) on amnesic patients with Korsakoff's syndrome, all argue for impaired spontaneous semantic structuring during encoding or retrieval as *one* cause of the memory deficit. Unfortunately, we do not know whether such an impairment was present in the patients of the semantic structuring transfer appropriate processing trainings reviewed above (Hildebrandt et al., 2006, 2011; Hildebrandt, Lanz, et al., 2007), and therefore we also do not know whether the treatment led to its remediation (but compare the following section on the loci of improvement in consolidation or compensation). Let us assume for convenience that the lack of spontaneous structuring *did* play a role in the impaired memory performance. If that was the case, then the training of semantic structuring and transfer appropriate processing deviates from teaching mnemonic strategies to memory-impaired patients in two aspects: first, by the (already mentioned) intrinsic involvement of both these processes in memory, and second, by focusing on a subprocess that is impaired. Mnemonic strategies, as used in the studies of Chiaravalloti et al. (2013, 2016), Kaschel et al. (2002), and B. A. Wilson (1987) are different in the fact that they should not be impaired. Indeed, the reason that they are taught to the patients is because they are *unimpaired* and therefore should be just as helpful as for healthy controls. Teaching mnemonics is *compensatory* in a strong sense of the word, because this approach tries to compensate for a deficit through the use of an intact function. Teaching semantic structuring and transfer appropriate processing may be at least partly restitutive, insofar as the memory impairment may be caused by an impairment of these processes.

The results of these studies on memory rehabilitation can be summarized as follows:

- The combination of the method of semantic structuring and transfer appropriate processing leads to significant improvement of memory functions, which is related to the intensity of the treatment and may reach moderate to strong effect sizes. The effect size difference was large for the group studies and low after converting the memory training into a PC software program (Figs. II.4.10 and II.4.11).

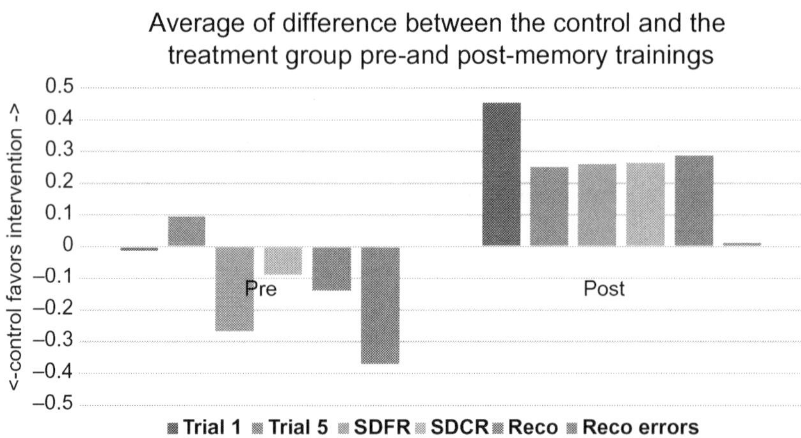

Fig. II.4.10 Calculation of the differences in CVLT performance for the control and treatment group before and after the training for four studies (Hildebrandt et al., 2006, 2011; Hildebrandt, Clausing, et al., 2007). *SDFR*, short delay free recall; *SDCR*, short delay cued recall; *reco*, recognition.

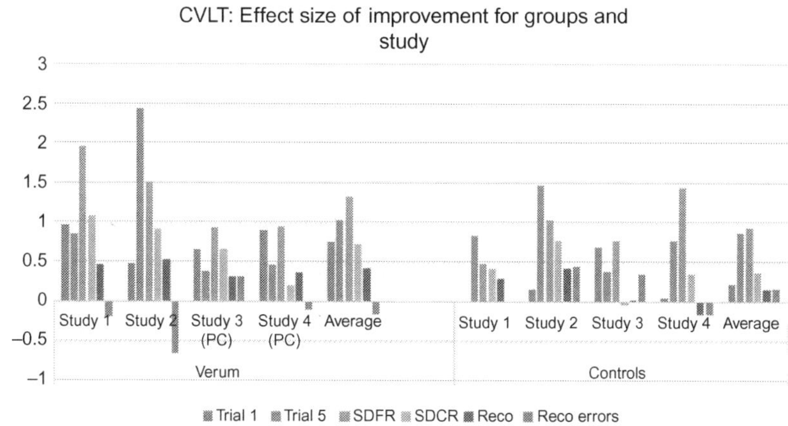

Fig. II.4.11 Post-hoc calculation of the effect sizes for the treatment and control groups (Hildebrandt et al., 2006, 2011; Hildebrandt, Clausing, et al., 2007). *SDFR*, short delay free recall; *SDCR*, short delay cued recall; *reco*, recognition.

- There are also some hints that these effects are clinically relevant because they transfer to text reproduction (which is an ecologically valid task). At least one study documented improvements in memory performance in activities of daily life, which were noted by the nurses who were blinded for group status. Thus, it seems to be fair to conclude that there is some evidence of transfer of this training to other cognitive functions and daily activities.
- In contrast to mnemonic strategies, the trained functions are probably quasi-automatic processes always involved in the memorization of information. Therefore, they are not additional tasks for the patients but rather something that they are always doing.
- Up to now, it is unclear if there is a specific subgroup of patients that profits from this training, because, for example, of a spontaneous nonuse of semantic structuring during encoding and retrieval. The results of Miotto et al. (2013) and Sheldon et al. (2012) indicate that *frontal* patients might especially benefit from using such a treatment approach. This would be an important piece of information for allocating the best treatment approach to different patient groups: Patients with mainly posterior lesions then should not be treated with this type of training. However, because there is no systematic study comparing patients with posterior (hippocampal) versus anterior (frontal) lesions and their response to treatment schedules (not only to a single experimental variation), we do not know whether this assumption is correct.

4.3 Effects of Semantic Structuring and Retrieval Appropriate Processing on Memory Performance: Internal Compensation or Restitution?

Up to now, it was purported that training of semantic structuring and transfer appropriate processing is effective because it improves encoding and retrieval. While this seems to be a reasonable hypothesis it may nevertheless be incorrect on a conceptual level. Reconsidering the proactive interference experiments of Cermak et al. (1974) conducted with Korsakoff's amnesics, one may also argue that there might be an inherent impairment in consolidating information. This is because the semantic structure of the word lists learned initially interferes with the lists learned later, and this effect is compounded over different presentations. The effect of the first presentation on the third or fourth presentation rules out that the deficit takes place during encoding.

There are other arguments that encoding is not the only locus of interference in amnesia. Baddeley and Wilson (2002), for example, studied two amnesic patients. The patients were severely impaired in word list learning but performed much better in text reproduction (at least in immediate recall, although not so much in delayed recall). The good performance in immediate text reproduction and the length of the text ruled out any explanation by means of the other components such as working memory (e.g., by rehearsal). Baddeley and Wilson (2002) responded to the observation that some amnesic patients can (1) recall text much better than word lists, and (2) perform an immediate recall of text better than in delayed recall by introducing an additional cognitive component: the episodic working memory buffer. This might have helped the patients to cope with interference during text reproduction, but not during word list learning.

Fig. II.4.12 is another modification of the Atkinson and Shiffrin (1968) model for memory. An episodic and a retrieval buffer have been added to the figure. At the same time, the transitions between encoding and consolidation and between consolidation and retrieval are now indicated as gradual processes. Instead of being rather encapsulated modules they now are steps of a

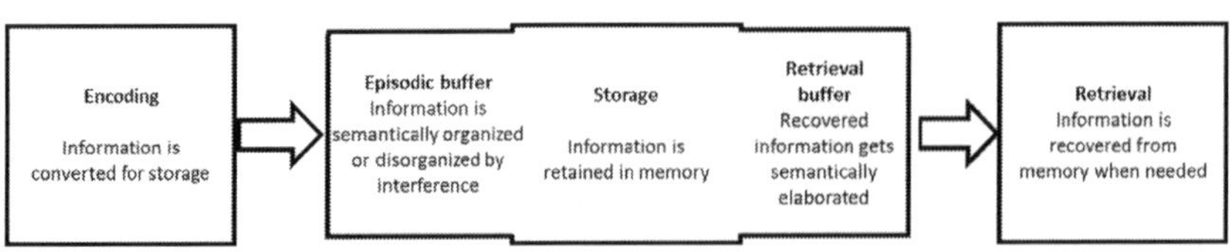

Fig. II.4.12 The Atkinson and Shiffrin (1968) model for memory augmented by the episodic buffer.

continuum of processes that lead to long-term storage. As shown by the impact of interference on memory performance, interference may work along the path of consolidating the information. Representing information in consciousness, (which was the "firewall" between encoding and consolidation and consolidation and retrieval), no longer plays a major role. This is because the episodic buffer stores information outside consciousness. Note that a model with gradual transition between encoding and consolidation also opens the possibility for some voluntary access to the consolidation process.

Dewar, Garcia, Cowan, and Della Sala (2009) and Dewar, Della Sala, Beschin, and Cowan (2010) published a series of experiments that also raises the question as to whether interference affects memory performance only during the encoding or retrieval processes. Table II.4.4 shows the general design of the experiments. Amnesic patients and healthy controls had to remember word lists. The experimental variation concerned the time *after* the presentation of the lists (Table II.4.4). The first group started the consolidation period with a three-minute easy attention task. Next was an unfilled period of 6 min where subjects would sit in a silent room without any task. The second group first had a three-minute unoccupied period in the silent room, followed by the three-minute simple attention task, and then ending with another three-minute, unoccupied rest. The third group started with a six-minute silent rest and then performed the attention task. The patients and controls were also tested without any attention task (fourth group).

Fig. II.4.13 shows the results of the four different conditions. The position of the attention task during the consolidation period showed a highly significant effect on the memory performance of the patients and only a small impact on that of the healthy controls. If the attention task was presented

Table II.4.4 The M. Dewar et al. (2009) design to test the effect of interference at different time-points in memory-impaired patients

<table>
<tr><td>Early</td><td>Word List</td><td>Immediate Recall</td><td>I</td><td colspan="2">Unfilled (6 min)</td><td>Delayed Recall</td></tr>
<tr><td>Mid</td><td>Word List</td><td>Immediate Recall</td><td>Unfilled (3 min)</td><td>I</td><td>Unfilled (3 min)</td><td>Delayed Recall</td></tr>
<tr><td>Late</td><td>Word List</td><td>Immediate Recall</td><td colspan="2">Unfilled (6 min)</td><td>I</td><td>Delayed Recall</td></tr>
<tr><td>Unfilled</td><td>Word List</td><td>Immediate Recall</td><td colspan="3">Unfilled (9 min)</td><td>Delayed Recall</td></tr>
</table>

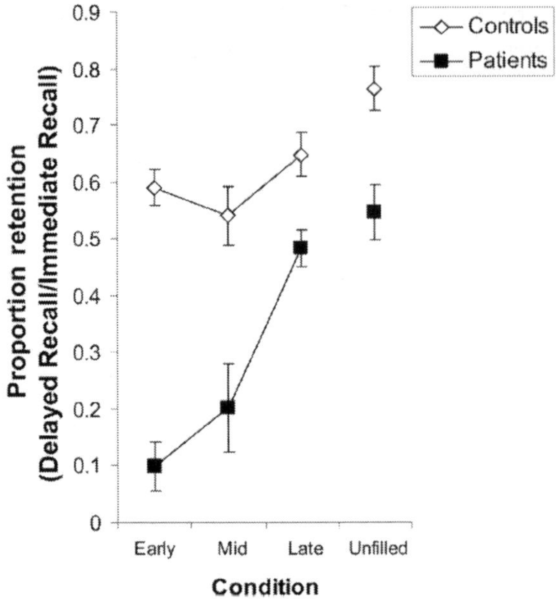

Figure 2.
Mean proportion retention (delayed recall divided by immediate recall) as a function of group and condition. Error bars show the standard error of the mean (S.E.M.)

Fig. II.4.13 Results of the M. Dewar et al. (2009) experiments.

after 6 min or not at all, the amnesic patients performed almost as well as the healthy controls. On the other hand, if the patients had the attention task at the beginning of the interval or after 3 min, the performance dropped to almost no memory at all. Of special interest is the group that started the task after a three-minute rest. The attention task still had a high impact on their performance. Therefore, interference seems to play a major role after finishing with encoding, i.e., already during the first consolidation phase.

Thus, recent knowledge about the effect of interference does not rule out the idea that it works during consolidation and after encoding is completed. Teaching the memory-impaired patients to cope with interference by using semantic structuring and transfer appropriate processing might also improve consolidation and not (only) encoding and retrieval. In this respect it is noteworthy that the patients of Dewar et al. (2009) suffered from moderate memory disorders but were not amnestic in a clinical sense of the word.

To analyze if semantic structuring and transfer appropriate process training have an impact on consolidation, and to answer the question of whether the improvements found were due to better encoding, consolidation, or retrieval performance, we reviewed the results of four studies (Hildebrandt, Clausing, et al., 2007). For this purpose, we tried to define specific parameters characteristically for these three processes and looked for changes after successful memory treatment. For encoding we analyzed the learning trials, the learning curve, and recognition; for retrieval the relation between free recall and recognition, the relation between cued and free recall, but also the improvement in word fluency; for consolidation, the relation between short-delay and long-delay free recall. Most of our parameters pointed in the direction that the training was successful in improving retrieval, while fewer results pointed in the direction of encoding, and none in the direction of consolidation. In this respect, it is also of interest that in the study by Hildebrandt et al. (2006), semantic structuring was only improved for delayed free recall and not for the learning trials. However, this review was performed retrospectively. None of the studies was designed to determine which of the three memory stages drives the improvements. Therefore, we set up a new study to answer this question. For this study (Hildebrandt et al., 2011, second study), we developed a new word list task. The word list was similar to that of the CVLT, but we introduced two new words at the third learning trial to analyze whether novelty would be detected by the patients before and after the treatment. Novelty detection is one of the major functions of the hippocampus and it served as a criterion as to whether the treatment improved the encoding process, i.e., if the patients would reproduce the new words more often. We also presented the word list after short-delay free recall another time, but it was now semantically structured to exclude any influence of a strategy shift on retrieval performance. Such a shift happens quite often with the CVLT, where the patients learn that the words can be semantically organized during the cued recall trial. They therefore perform better on long-delay free recall than on short-delay free recall. The difference between the number of words reproduced after a semantically structured presentation of the word list and after a delay period of an additional 20 min was defined as an indication for consolidation. Retrieval performance was measured by the difference between cued recall and free recall, or recognition and free recall. The study was designed as a RCT, and we compared the performance of the usual treatment at the rehab unit with the PC-based variant of the semantic structuring and transfer appropriate processing training. In contrast to a former study (Hildebrandt et al., 2011, experiment 1), this time the patients trained alone at different PCs but were in the same room with other participants. Thus, there was no therapeutic interaction between the therapist and the patients.

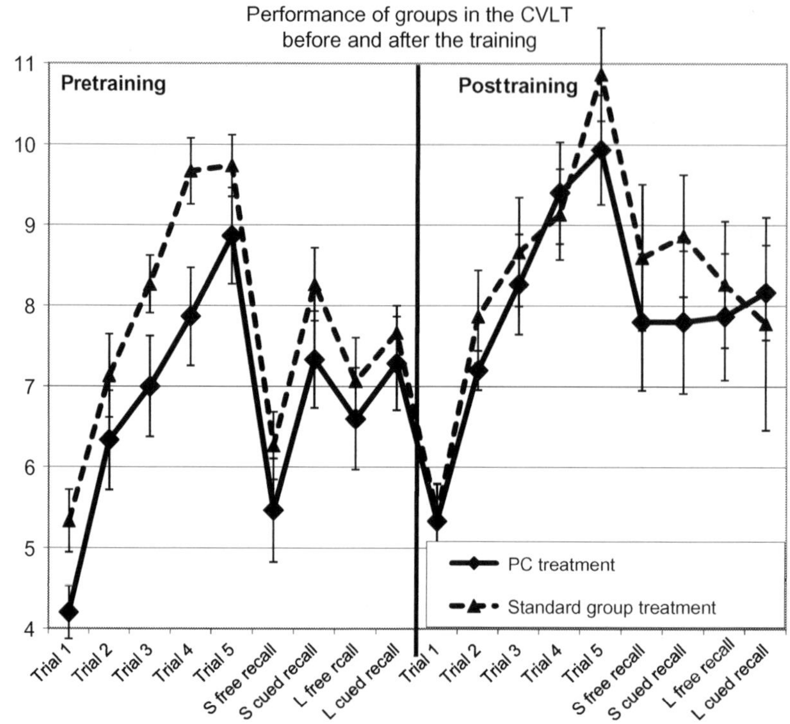

Fig. II.4.14 Results of second study by Hildebrandt et al. (2011).

In contrast to the previous studies, the investigation failed to show a superior improvement of the semantic structuring and transfer appropriate processing group. However (in absolute terms), during the learning trials the latter group improved more than the treatment as usual group (Fig. II.4.14).

Therefore, it was not possible to distinguish between the two treatments. Collapsing the data of both groups, performance changes were limited to parameters that we defined as being characteristic for retrieval performance (Table II.4.5). Thus, in contrast to the studies of the group of Dewar et al. (2009, 2010), we did not find any indications for an improvement in consolidation.

However, it may be the case that we missed effects on encoding and consolidation, just because the intervention was not successful. Thus, we had to rely on all participants and were not able to really test the assumption whether the semantic structuring transfer appropriate processing training showed such kinds of effects.

Table II.4.5 Does memory training improve encoding, consolidation, or retrieval? Results of the second study of Hildebrandt et al. (2011)

Memory stage	Scores to analyze memory stages by the ANOVA for repeated measurement	Results (significance level met in ANOVA for repeated measurement (yes/no))
Encoding		
	CVLT & DRMAD: First to last learning trial	No
	DRMAD: reproducing novel words from the third and fourth learning trial	No
Consolidation		
	CVLT & DRMAD: Short delay to long delay free recall	No
Retrieval		
	CVLT & DRMAD: Long delay free recall to recognition	Yes
	CVLT: Effect of semantic cueing	Yes
	DRMAD: Short delay free recall to free recall after structured presentation	Yes

Abbreviations: *CVLT*, California Verbal Learning Test; *DRMAD*, Delayed Recall Memory Test for Alzheimer's Disease.

We can conclude that the treatments discussed up to now probably do not improve consolidation but instead just encoding and retrieval. How important is this finding for clinical memory rehabilitation? In other terms, how many patients suffer from a consolidation deficit (fast forgetting) and how many from an encoding or retrieval deficit? There are no clear-cut figures to answer this question, but nevertheless there are some important hints. Piras, Borella, Incoccia, and Carlesimo (2011) reviewed studies on memory rehabilitation and tried to distinguish between different groups of patients. They found that most of the patients suffered from mild to moderate impairments, and they identified only 5 studies with a total of 37 purely amnesic patients. Fast forgetting is a specific problem of patients with hippocampal lesions or lesions of the Papez circuit. Most stroke patients do not suffer from such lesions, because only a small number of strokes involve the medio-temporal branches of the posterior cerebral artery or the anterior and medial structures of the thalamus. Re-evaluating the included patients of two of our studies (Richter, Modden, Eling, & Hildebrandt, 2015, 2018) showed that only 19% of all patients might have suffered a (unilateral) hippocampal lesion. Moreover, comparing the results of these patients with that of the other patients revealed no consolidation deficit during the delay period between short- and long-term free recall. Rather, it showed an accelerated loss of information when comparing the last learning trial and short-term free recall, which is usually explained by a higher susceptibility to retroactive interference (Griffiths et al., 2005).

TBI is another major cause for organic memory disorders. DeLuca, Schultheis, Madigan, Christodoulou, and Averill (2000) analyzed explicitly whether TBI patients suffer from consolidation deficits or deficits in encoding and retrieving information. They tried to match the performance of the patients during encoding with that of healthy controls by providing the patients with more time and trials for learning. Then comparing the forgetting curves of the patients with that of the healthy controls for two different delays, DeLuca et al. (2000) did not find any difference between the groups. Therefore, they argued that the main problem of memory-impaired TBI patients is the initial acquisition of verbal information rather than a compromised consolidation, i.e., a similar pattern as we found in stroke patients. Similar results comparing initial learning and consolidation performance were documented by Chiaravalloti et al. (2014) for patients with Parkinson's disease and by DeLuca, Leavitt, Chiaravalloti, and Wylie (2013) for patients with multiple sclerosis. It can be concluded that many (if not most) of the mildly to moderately impaired patients suffer from deficits in encoding and retrieval and not (so much) from consolidation deficits. Thus, although the ultimate goal of memory rehabilitation is to develop tools for different (that is, any type of) memory impairments, focusing on encoding and retrieval seems to be correct for treating the majority of patients. The findings of Chiaravalloti et al. (2014) and DeLuca et al. (2000, 2013) also show that the highly influential assumption of B. A. Wilson (1987) that memory deficits cannot be cured was an incorrect generalization of findings and were based on the famous patient HM. For the majority of patients with memory deficits there might be a way of "curing" the deficits (at least somewhat), but there is also a group of patients with severe deficit, where any type of direct training might be senseless and the statement of B. A. Wilson (1987) holds.

5 IMPROVING MEMORY PERFORMANCE THROUGH WORKING MEMORY TRAINING

5.1 Is Working Memory Training Effective?

Spencer-Smith and Klingberg (2015) published a very influential review on the effect of working memory training on working memory tests (near transfer) and daily life performance (clinical scale scores). This review partly summarized the work of Klingberg's lab, focusing on working memory training in young attention deficit hyperactivity disorder (ADHD) patients, in stroke patients, and in older adults. Fig. II.5.1 shows the result of the review for daily life performance and Fig. II.5.2 shows the results for near and far transfer. There was a moderate effect on daily life performance (−0.47 in favor of an improvement) and a higher effect on visuospatial memory, but no effect on verbal working. In this respect, it should be mentioned that the training modules that most of these studies were using focused on visuospatial working memory. Furthermore, there are some arguments for visuospatial working memory showing more plasticity and meaning for daily life than verbal working memory (especially if conceptualized such as relaying a series of numbers forwards and backwards).

Daily life measures

	Cogmed			Control				Std. mean difference
Study or subgroup	Mean	SD	Total	Mean	SD	Total	Weight	IV, Fixed, 95% CI
Beck 2010	5.72	2.92	27	7.5	1.79	24	10.3%	−0.71 [−1.28, −0.15]
Brehmer 2012a	25.66	12.37	29	31.69	12.39	26	11.6%	−0.48 [−1.02, 006]
Brehmer 2012b	31.86	9.77	26	38.79	19.11	19	9.3%	−0.47 [−1.07, 0.13]
Chacko 2013	−16.5	5.84	38	−14.24	6.06	36	15.8%	−0.38 [−0.84, 0.08]
Egeland 2013	15	5.6	33	16.2	6.2	34	14.5%	−0.20 [−0.68, 0.28]
Green 2012	67	12.2	12	70.9	8.3	14	5.5%	−0.37 [−1.15, 0.41]
Gropper unpublished	48.75	17.72	39	55.65	15.11	23	12.3%	−0.41 [−0.93, 0.12]
Grunewaldt 2013	−6.9	5.3	9	−5.1	4.5	11	4.3%	−0.35 [−1.24, 0.54]
Hardy 2013	60.8	15.42	11	74	8.08	6	3.0%	−0.93 [−1.99, 012]
Klingeberg 2005	13	6.5	17	15.5	7.2	19	7.7%	−0.36 [−1.02, 0.30]
Raoughan 2011	12.57	11.12	7	28.88	11.69	8	2.5%	−1.34 [−250, −0.18]
Westerberg 2007	29.2	12.1	9	43	13.8	9	3.4%	−1.01 [−2.01, −0.02]
Total			257			229	100%	−0.47 [−0.67, −0.29]

A negative value for the std. mean difference denotes *increased* working memory performance

Fig. II.5.1 Working memory training: effects on daily life measures (Spencer-Smith & Klingberg, 2015).

Visuospatial working memory

	Cogmed			Control				Std. Mean difference
Study or subgroup	Mean	SD	Total	Mean	SD	Total	Weight	IV, Fixed, 95% CI
Brehmer 2012a	7.14	0.97	29	5.83	0.42	26	13.1%	1.70 [1.07, 2.32]
Brehmer 2012b	5.69	0.65	26	5.26	0.67	19	13.8%	0.64 [0.03, 1.25]
Chacko 2013	107	19.21	37	99.97	17.15	36	23.8%	0.38 [−0.08, 0.84]
Gropper unpublished	0.185	0.96	39	−0.205	1.18	23	18.9%	0.37 [−0.15, 0.89]
Grunewaldt 2013	−2.3	2	9	−2.7	1.6	11	6.%	0.21 [−0.67, 1.10]
Hardy 2013	12.4	4.57	10	10.4	2.3	7	5.3%	0.5 [−0.49, 1.48]
Klingeberg 2005	5.22	1.01	20	4.28	0.81	24	12.7%	1.02 [0.38, 1.65]
Westerberg 2007	6.2	1	9	5.7	1.8	9	5.6%	0.33 [−0.60, 1.26]
Total			179			155	100%	0.66 [0.43, 0.89]

A positive value for the std. mean difference denotes *increased* working memory performance

Verbal working memory

	Cogmed			Control				Std. Mean difference
Study or subgroup	Mean	SD	Total	Mean	SD	Total	Weight	IV, Fixed, 95% CI
Brehmer 2012a	6.76	1.66	29	5.67	1.41	26	16.2%	0.69 [0.15, 1.24]
Brehmer 2012b	5.94	1.58	26	5.5	1.43	19	13.7%	0.28 [−0.31, 0.88]
Chacko 2013	104.4	11.9	37	103.4	14.56	36	23.0%	0.07 [−0.39, 0.53]
Gropper unpublished	10.2	2.99	39	8	2.77	23	17.0%	0.75 [0.22, 1.28]
Grunewaldt 2013	1.9	1.2	9	1.1	1.3	11	5.9%	0.61 [−0.30, 1.52]
Hardy 2013	9.9	3.75	10	9.8	2.56	7	5.2%	0.03 [−0.94, 0.99]
Klingeberg 2005	4.1	0.89	20	3.73	0.73	24	13.4%	0.43 [−0.17, 1.03]
Westerberg 2007	5.8	1	9	5.7	0.9	9	5.7%	0.10[−0.82, 1.02]
Total			179			155	100%	0.40 [0.18, 0.62]

A positive value for the std. mean difference denotes *decreased* working memory performance

Fig. II.5.2 Working memory training: effects on neuropsychological test results (Spencer-Smith & Klingberg, 2015).

The results of the review were quite convincing. Moreover, other groups showed that working memory can also be trained in older adults (Borella, Carretti, Riboldi, & De Beni, 2010; Li et al., 2008) and that such training influences the dopamine concentration in the brain (Klingberg, 2010). In older adults the effects of this type of training can be demonstrated with the help of diffusion tensor imaging (Lövden et al., 2010), arguing for its effect on the remyelination of axons.

The effects of working memory training in younger adults and transfer effects of this training on other neuropsychological tests and daily life performance in *healthy subjects* are still controversial (Soveri, Antfolk, Karlsson, Salo, and Laine, 2017; Soveri, Karlsson, Waris, Gronholm-Nyman, and

Laine, 2017). Nevertheless, the review of Spencer-Smith and Klingberg (2015) argued that this type of training might be helpful for *patients*. However, most of the studies cited in the review dealt with psychiatric patients and not with neurological rehabilitation patients, i.e., suffering from brain damage acquired as adults. Even more importantly, many of the included studies compared the effects of working memory training to a no-training group, i.e., did not fulfill the criteria of double or triple blinding. Therefore, it is not clear whether motivation due to involvement in the treatment played a major role for the positive results of the studies. Moreover, the review did not concern episodic memory functions but working memory. Although working memory may play a role in many cognitive functions, its meaning for *episodic* memory performance is not completely settled.

There are (at least) three studies concerning the effects of working memory training in patients with acquired brain damage (Akerlund, Esbjornsson, Sunnerhagen, & Bjorkdahl, 2013; Lundqvist, Grundstrom, Samuelsson, & Ronnberg, 2010; Westerberg et al., 2007; see Spreij, Visser-Meily, van Heugten, & Nijboer, 2014 for a comprehensive review ignoring some important EBM criteria) and only one of them was included in the Spencer-Smith and Klingberg (2015) review. None of these three studies were a placebo-controlled trial, because all compared intervention to a passive (waiting) control group. There was no blinding of the assessor in the Westerberg et al. (2007) study, and no blinding at all in the two other studies. The Westerberg et al. (2007) study reports the concealment, while the other two studies do not mention whether there was any concealment at all. All three studies found an improvement in working memory tests (near transfer). There was not a far transfer effect on other neuropsychological functions. The experimental (but not the passive control) group in the Westerberg et al. (2007) study reported a decrease of cognitive failures in the "Cognitive failure questionnaire". Unfortunately, this type of subjective improvement in the patients receiving the treatment compared to those not receiving treatment may be explained by a simple placebo effect. To summarize, working memory training alone seems to lead to improvements in working memory performance of patients with acquired brain damage. Far transfer effect in this patient group, especially in regard to episodic memory performance, has yet to be demonstrated.

Although gains in untrained tests of working memory are commonly reported as an effect of working memory training, the causes for these gains are unknown. There are two different explanations for the improvements. The first argues that the working memory *capacity* is increased by the training. Therefore, more information can be held in working memory, and at the same time this leads to improved performance. The other explanation argues that the training leads to the development of *cognitive strategies* adapted to the task's demands. Confronted with untrained, but similar, tasks the subjects will transfer the newly developed strategies and therefore will show gains in performance. Dunning and Holmes (2014) investigated whether working memory training improves performance in untrained working memory tasks. They compared three groups of young healthy subjects, one receiving an adaptive working memory training, the second a nonadaptive low-level working memory training, and the third group was a passive control group. After the training, the participants were asked if they developed strategies to perform the training tasks. The answers were evaluated and categorized into different strategies such as rehearsal, grouping, visualization, imagery, inhibition of irrelevant information, chunking, and chaining. The group that received adaptive working memory training showed significant improvement in untrained tasks such as visuospatial short-term memory and verbal working memory, whereas the two control groups did not differ in their performance. Members of these groups reported significantly more often that they developed a strategy to solve the untrained task, and the most dominant strategy was *semantic grouping*. A direct comparison of the performance of participants who answered that they used the method of semantic grouping after the training with those who did not report an increase of its use was not significant. Dunning and Holmes (2014) argued that this can be explained by the low number of participants remaining in the statistical analysis for such a comparison. Thus, in young healthy subjects working memory training alone induces a strategy change towards semantic grouping. This type of strategy change is the central goal of treatment approaches such as semantic structuring and transfer-appropriate processing training.

Recent results from functional brain imaging underline that the harsh distinction between working memory and episodic memory, which is one aspect of the Atkinson and Shiffrin (1968) model, may be an exaggerated separation. Rose, Craik, and Buchsbaum (2015), for example, analyzed the performance of healthy controls in working memory tasks for three conditions: active rehearsal, distraction by easy mathematical tasks, and distraction by hard mathematical tasks. The content to be kept in working memory was divided into shallow and deep (semantic) information. Rose et al. (2015) found that semantic information could be represented in working memory more efficiently than information that was encoded on a shallow level. Moreover, the results from functional brain imaging showed that the putative storage of the information during the working memory task was the same as during episodic memory tasks. This would argue that working memory and episodic memory share similar processes (e.g., deep and shallow encoding being similarly efficient in both memory functions) and similar storage structures, with the main characteristic of working memory being that it is a consciously controlled process (for a review of such a model for working memory compare D'Esposito & Postle, 2015).

5.2 Working Memory Training to Improve Episodic Memory

The lack of reports on far transfer of working memory training to episodic memory performance is surprising given the seemingly intrinsic relation between episodic memory and working memory. Although both functions can be dissociated in patients with different types of brain damage (as mentioned above), they may share the same processes and storage structures. There is a lot of empirical and conceptual evidence that working memory plays a role in learning und recall, because of its prominent role during encoding and retrieval. To give only two examples: Duff, Schoenberg, Scott, and Adams (2005) analyzed the performance of 212 patients referred for a neuropsychological evaluation in executive tasks and in learning tests. They found that there is a strong correlation between both functions, suggesting that performance in both may be causally related. On a conceptual level, Baddeley and Wilson (2002) felt the need to introduce the episodic memory buffer to allow for some interaction between working memory and learning in their model.

Fig. II.5.3, which is variation of the Atkinson and Shiffrin (1968) model, tries to capture the strategic role working memory may play for encoding and retrieval.

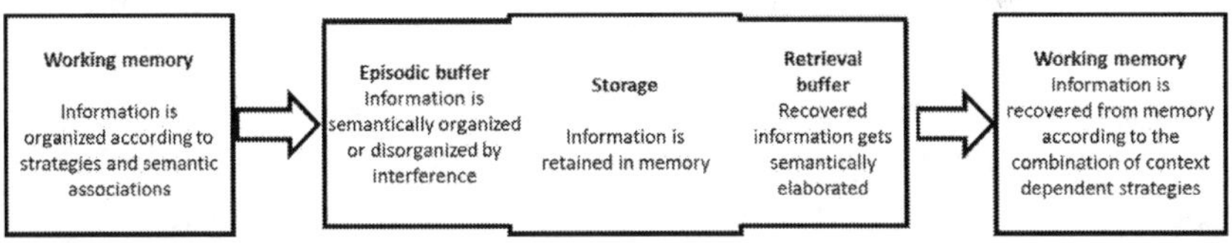

Fig. II.5.3 The augmented Atkinson and Shiffrin (1968) model and the putative role of working memory for encoding and retrieval.

The model shows that working memory may have a major impact on semantic structuring, especially in verbal information, where the integration of the meaning must be performed by working memory. It also is an essential component of strategic search during retrieval and it has been argued that the effects of semantic structuring and appropriate transfer processing training rely on improved retrieval processes. Why then did working memory training not show a positive impact on verbal learning and recall in the studies involving patients with acquired brain damage? One answer to this question is that only the Westerberg et al. (2007) study measured verbal learning, while the two other studies of Akerlund et al. (2013) and Lundqvist et al. (2010) focused on working memory, inhibition, and attention. However, from an EBM viewpoint the Westerberg et al. (2007) study is superior compared to the two other studies. A second answer might be that generalization and transfer should be an integral component of the treatment schedule. Because none of the three studies included any type of generalization practice, the patients might have failed to show far-transfer effects. There is yet another explanation as to why the three studies did not show such a generalization. Two focused exclusively on visuospatial working memory function using the Cogmed QM® training module developed by the Klingberg group. Only the Lundqvist et al. (2010) study also involved verbal working memory tasks. In this respect, it should be noted that the Spencer-Smith and Klingberg (2015) review showed that verbal working memory did not necessarily improve after training with tasks from the Cogmed QM®. Therefore, training with Cogmed QM® may lead to improvement in visuospatial working memory and cognitive functions related to right frontal areas (reasoning, calculation, etc.) but not to functions related to the left hemisphere such as language and verbal memory.

The findings of Dunning and Holmes (2014) state that generalization of working memory training prominently relies on the spontaneous transfer of strategies, such as using semantic grouping for new tasks. Therefore, a combination of working memory training and semantic structuring should offer an ideal possibility for the rehabilitation of memory deficits. The working memory training would predispose the patients to rely on semantic grouping, and the additional training of semantic structuring would help the patients to rely on this type of strategy for encoding and retrieval.

To analyze whether verbal memory performance can be improved by such a combination of working memory and semantic structuring, Richter et al. (2015) set up a RCT trial with envelope-based concealment. The assessor was blinded at t2 for the group status of the patients. A standard memory training was provided to the active control group. Therefore, the patients of both groups were unaware of their treatment status. Patients of both groups (each group with $n = 18$ patients) received exactly the same number of treatment hours (9 h of treatment over 3 weeks). Inclusion criteria were an impaired learning performance (percentile rank < 16) and/or an impaired short-term free recall performance in the CVLT (percentile rank < 16). Severely memory-impaired patients that recalled <3 words at short-term free recall were excluded from the study.

The experimental group received a semantic structuring training without a transfer-appropriate processing component. They also trained their working memory with WoMe, a software tool of the RehaCom battery (Hasomed®). The WoMe training is a computer-based card game. The task is to reproduce an increasing number of cards in the same or reversed order or to memorize only cards of a nominated suit. Moreover, at higher levels attention is drawn away through the presentation of different distractors. Semantic structuring was trained by the presentation of lists of words stemming from two different categories or action sets. The patients then had to find and propose headings to organize the lists. The lists were also divided in easy, moderate, and hard exemplars, and the presentation was adapted to the performance of the patients. Moreover, different types of word-fluency tasks were used to train the ability of patients to move around in semantic networks and to improve semantic structuring. Last, but not least, patients in the experimental group received no specific memory training and were taught no mnemonic strategies. Instead, they underwent only verbal working memory training (WoMe) and training of semantic structuring. The active control group received training for 3 weeks, 3 h a week (the rehabilitation unit standard treatment). The active control group practised memory tasks, were informed about strategies to improve memory performance, and played games that involved memory performance.

The combination of WoMe and semantic structuring led to a significantly higher improvement in working memory, in word fluency, and in prospective memory performance (Fig. II.5.4). The standardized EF (i.e., the test gains of the treatment group reduced for the improvement in the active control group) for working memory was <0.3, for word fluency approximately 0.8, and for prospective memory approximately 1.

Regarding immediate and delayed memory, Richter et al. (2018) used two composite scores that involved the results from the CVLT, as well as for text reproduction of the Rivermead behavioral memory test. There was no difference between the groups: The amount of improvement was quite similar (Fig. II.5.5). This argues that the treatment involving semantic structuring and WoMe had a similar effect on memory performance as its specific training in the active control group, and can be regarded as an indication of a far-transfer effect (additional to the higher improvement in prospective memory).

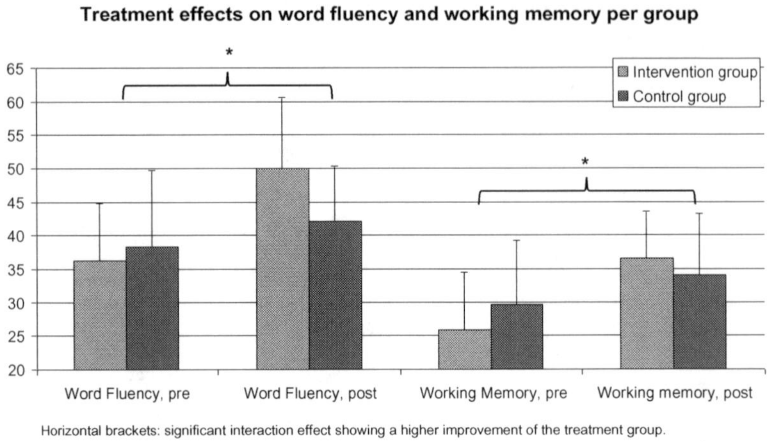

Fig. II.5.4 Results of the Richter et al. (2015) study on near and far transfer.

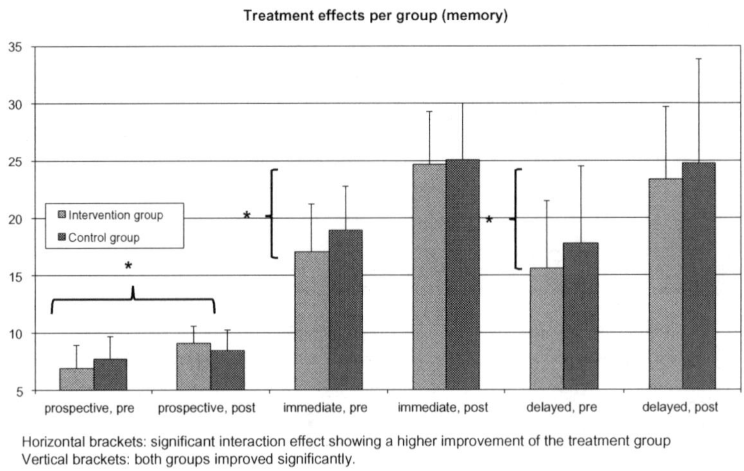

Fig. II.5.5 Results of the Richter et al. (2015) study on improvement in learning.

Both groups did not differ in the cognitive failure questionnaire, neither before nor after the treatment (however, recall the notoriously weak meaning of subjective reports for the evaluation of memory training). The intragroup effect size of the improvement in immediate and delayed memory was quite high: The effect size in both groups was at least 1, in some instances higher than 1.5. There is some indication that the high effect size in both groups is partly due to spontaneous recovery: Time since lesion was about 2 months with a standard deviation of more than 2 months. This argues that there were a few patients who had been living with their brain damage for a long time and there was a larger group in an early post-acute stage. Both groups did not improve in attention performance during the training. This argues that the treatment effects cannot be explained by a non-specific global remission effect.

Thus, the study of Richter et al. (2015) shows that working memory training combined with semantic structuring training improves verbal learning and recall as good as a specific memory training, but additionally increase working memory, prospective memory, and word fluency. The level of the study is Ib, but the included patient number is small and there is no follow-up. The study clearly shows far-transfer effects of the training cognitive training, but there is a lack of tasks or questionnaires assessing the impact on everyday memory performances rated by relatives unaware about the specific treatment of the patients.

6 MEMORY REHABILITATION AS IMPROVEMENT OF CONSOLIDATION

6.1 Dissociating Consolidation: The Dual Process Theory of the Memory Trace

Up to now, the central goal of memory rehabilitation was to increase free recall of learned information. This can be termed "hits", i.e., the number of correct items reproduced after a delay. In some patients with memory impairment, a low number of hits is not the only or even the most prominent problem for their relatives. Patients with moderate to severe memory impairment tend to be able to repeat what has just been said to them. Their major impairment seems to concern the appropriateness of retrieval in a given situation, not the inability to retrieve the information per se. Patients may tend to repeat what they have already said over and over again and it is this behavior which produces major discomfort on the side of the relatives. Schacter, Verfaellie, Anes, and Racine (1998) and Budson et al. (2002) published papers on the memory performance of acquired memory-impaired patients and in patients with Alzheimer's disease. In both articles, the Deese/Roediger–McDermott paradigm was used. Participants were shown a study list of words which were highly associated with another word that was not part of the study list. For example (Fig. II.6.1), many words from the study list were associated with "sweet", but the word sweet was not shown.

During the recognition test phase, the participants had to decide if the words shown to them were elements of the study list or not. The Deese/Roediger/McDermott paradigm, which is the common term for this type of learning and recognition experiments, leads to a high number of *false recognitions*, even in healthy controls. The probability for a lure word such as sweet to be recognized as "correct" is as high as the probability of recognizing of an actually presented word (i.e., a "hit"). However, healthy controls can increasingly suppress such false recognitions when studying the list several times.

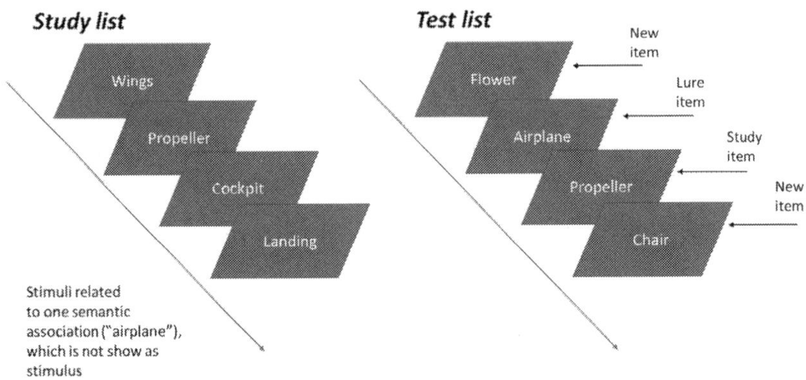

Fig. II.6.1 The Deese/Roediger–McDermott paradigm (Budson et al., 2002).

Schacter et al. (1998) and Budson et al. (2002) showed that the opposite is true for many patients suffering from amnesia. In these patients, the lures are learned in the same way as if they had been presented (Figs. II.6.2 and II.6.3). Obviously, amnesic patients are not able to discriminate between semantically associated and really presented words. Similar (but not so pronounced) findings were reported for other amnesic groups such as patients with Huntington's disease using the CVLT, which also showed problems to suppress related unrelated lures during the recognition trial (Delis et al., 2005). This inability to suppress internally generated representations as irrelevant brings to mind the problem of some memory-impaired patients being able to suppress what they have already said several times. In both cases the common problem seems to be a loss of context of information instead of the information itself.

Why do many amnesic patients show an incorrect learning of lures as correct items, despite studying the same list several times? Why are they not suppressed? Schacter et al. (1998) and Budson et al. (2002) explained their findings using the dual trace model of memory. According to this

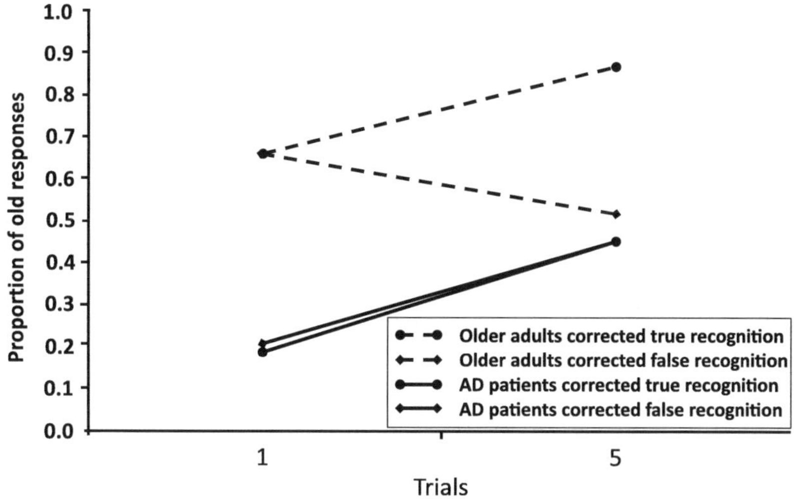

Fig. II.6.2 Corrected true and false recognition obtained by subtracting the proportion of "old" responses to unrelated items from the proportion of "old" responses to studied words and related lures, respectively, in patients with Alzheimer's disease and older adults as a function of one versus five study sessions (Budson et al., 2002).

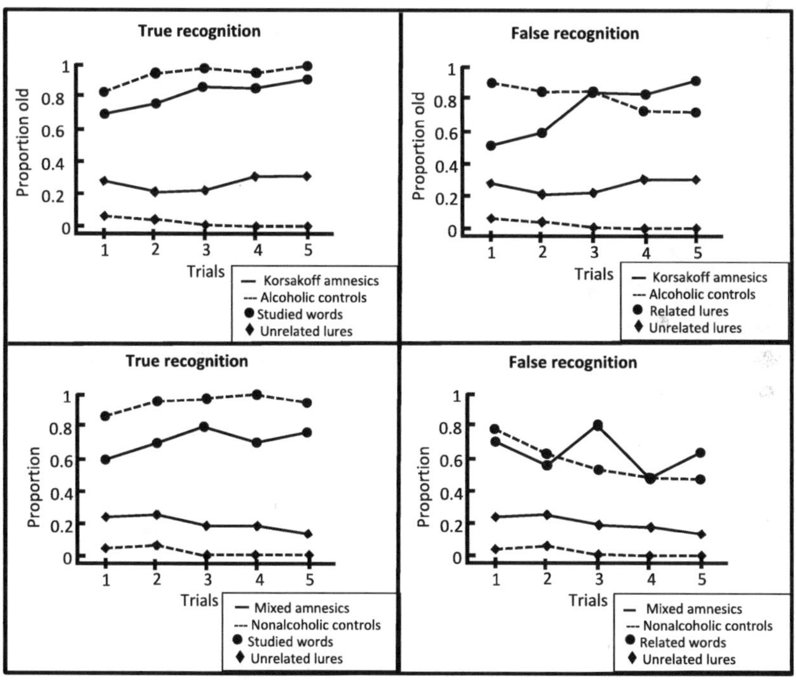

Proportions of old responses to studied words (Panels A and C), related lures (Panels B and D), and unrelated lures (Panels A-D) in the two subgroups of amnesics (Korsakoff and mixed) and their respective control groups (alcoholic and nonalcoholic) as a function of studytest trial. Korsakoff and mixed amnesics showed similarly impaired true recognition. However, Korsakoff patients showed increasing false recognition across trials, whereas mixed amnesics showed a fluctuating pattern across trials.

Fig. II.6.3 Results of the Deese/Roediger–McDermott paradigm in Alzheimer's patients and older adults (Schacter et al., 1998).

model, learning takes place through two separate mechanisms: One mechanism deals with automatic and unconscious learning, and this mechanism works based on repetition and knowing. The other mechanism concerns conscious remembering. Subjects try to "replay" the situation and its context and then judge if the critical aspects were present or not. Thus, conscious "replay" of the experienced information not only concerns the information itself, but also when, where, and how it has been experienced (in other words, the context). Remembering through the first mechanism is usually called *remembering by familiarity*, while remembering with the second mechanism is called *remembering by recollection*. Supposing that amnesic patients suffer from a loss of recollection, they will be unable to use the context to distinguish if the specific word has been on the list, or if it was just a memory trace that became activated due to its associations. Therefore, they will tend to learn associations strongly related to the presented items as if they were on the list. Unimpaired healthy controls will be able to use context memory to suppress such false recognitions. The increased number of false positives in amnesic patients is then a result of unconsciously forming semantic associations during the study phase plus a deficit in remembering the context (mere association or studied word). A similar mechanistic explanation may hold for the repetition of sentences by amnestic patients: each repetition increases the memory strength of the sentence and therefore leads to its repetition. Additionally, the loss of the memory of the context ("I have already said this to this person before") hampers the suppression of saying it again.

There is still a debate on whether the distinction between these two mechanisms is really a fundamental one, but it seems to be fair to argue that most of the available data endorse the "dual trace" or Remember/Know model of memory. Recollection and familiarity are related to different neuroanatomical structures. Recollection is associated with proper function of the fornix, mammillary bodies, and the anterior nucleus of the thalamus, as nodes of the Papez circuit. The most important neuroanatomical structure for recollection is the hippocampus which is related to context-specific memory consolidation. Familiarity is based on activity in various brain areas and on distributed cortical networks, with the parietal lobe, prefrontal areas, and the parahippocampus being the most important.

Patients with isolated bilateral lesions of the hippocampus are very rare. Nevertheless, evidence from the few patients with probable isolated lesions in the hippocampus and indirect evidence from patients with interruption of the Papez circuit argues for the view that recollection is based on the proper functioning of these structures. Vann et al. (2009), for example, used different methods to distinguish between remembering by familiarity ("knowing") and remembering by recollection ("remembering"). This study compared patients suffering from a colloidal cyst, a type of tumor that grows next to the columns of the fornix, i.e., at the anterior wall of the third ventricle. Colloidal cysts tend to squeeze the fornix and this squeezing may lead to degeneration of the mammillary bodies, one major effector structure of the fornix. Comparing patients with small ("atrophic") and large ("non-atrophic") mammillary bodies, Vann et al. could show that they differed systematically in recognition performance. Even more interesting was that this difference was due to a systematic loss of *recollection* performance in the patients with small mammillary bodies, whereas both groups did not differ in familiarity (Fig. II.6.4). Because they used a variety of different methods to

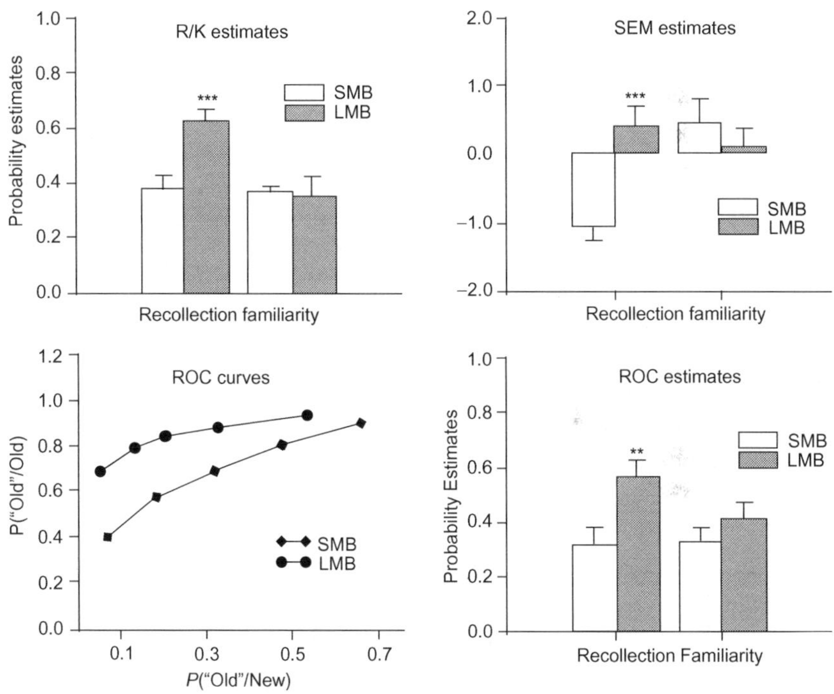

Fig. II.6.4 Derived probability estimates of "recollection" and "familiarity" for the small mammillary bodies (SMB) and large mammillary bodies (LMB) groups, drawn from a cohort of postsurgery colloid cyst patients. Data are represented in histograms as means ± standard error of the mean. (A) Estimates from the Remember/Know (R/K) procedure. (B) Estimates from structural equation modeling (SEM). (C) Receiver operating characteristic (ROC) curve. (D) Estimates from receiver operation characteristic (ROC procedure). $^{**}P < 0.01$; $^{***}P < 0.005$ (Vann et al., 2009).

differentiate between familiarity and recollection, and all these methods showed a similar result, the investigation offered strong evidence that each recognition performance is the product of two separate and independent processes.

Several years ago, we (Hildebrandt, Muller, Bussmann-Mork, Goebel, & Eilers, 2001) studied a patient who underwent the removal of a craniopharyngioma. A high solution structural MRI scan performed after the removal documented that the patient suffered from bilateral shrunken mammillary bodies and an infarct of the left mammillary body, consequences that were probably related to the tumor growth. The patient showed mild executive impairments, almost intact levels of learning and recognition, and impaired delayed free recall. A more detailed investigation revealed that the patient had enormous difficulty in distinguishing between information of different recency and of different sources in forced choice recognition. Thus, although familiarity performance was intact (no relevant impairment in recognition), he was unable to remember the context of the information.

6.2 Is it Possible to Improve Recollection? The Repetition Lag Procedure

The distinction between familiarity and recollection, or knowing and remembering, is based on two psychological states or processes. One may also stretch the "content" side of this distinction. The content of familiarity is then the "item" to be remembered, the content of recollection is the context (or relational embeddings) in which the item was experienced. Some authors therefore distinguish between item-specific memory and context-memory instead of familiarity and recollection. From this perspective, it seems to be reasonable to assume that the context might help to remember the items to be learned. Smith (1984) published a paper where he varied the situational context during learning. He analyzed whether instructing the participants to use such contextual information to enhance recall would improve their memory performance. His experiment had three groups. Subjects of all groups were given three word lists to study in one or three rooms. All subjects were given a free-recall test in a new room, and half were asked to use remembered environmental context information to facilitate word memory. Learning the three word lists in one room and being cued to imagine this previous room during recall improved performance by eight words (44% of the uninstructed groups' mean). Subjects of three-room learning condition profited from environmental context cues less than participants of the one-room learning condition, and recalled only a little more than the uninstructed group. A more detailed comparison of the proportion of recalled words from the three word lists showed that the group learning them in three different rooms profited from the cue to imagine the first room as much as the group with only one room. However, this beneficial effect started to decline for the second word list (= second room) and declined even more for the third word list (= third room). Smith (1984) concluded that contextual information could be used systematically for cueing item-specific memory performance, but also that multiple contexts (e.g., increasing number of rooms) could interfere with each other, either proactively or retroactively.

In terms of memory rehabilitation S. M. Smith (1984) could show that it might be possible to train patients not only to use specific strategies during encoding, but that it is also possible to inform them about strategies to be used during retrieval. As mentioned in his paper, a 44% difference in relation to the uninstructed group is an impressive success. Still, this approach obviously presupposes different encoding situations that may be used for superior retrieval of information. Is it possible to train recollection without definite contextual cues? Starting with the remember/know model for memory performance, Jennings, Webster, Kleykamp, and Dagenbach (2005) used the repetition lag procedure to increase the use of the recollection component and thus to improve contextual consolidation of information. The repetition lag procedure is composed of two phases: a learning part (studying a word list) and a recognition part (judging if the presented words have been part of the study list) (Fig. II.6.5). During the recognition phase, the newly

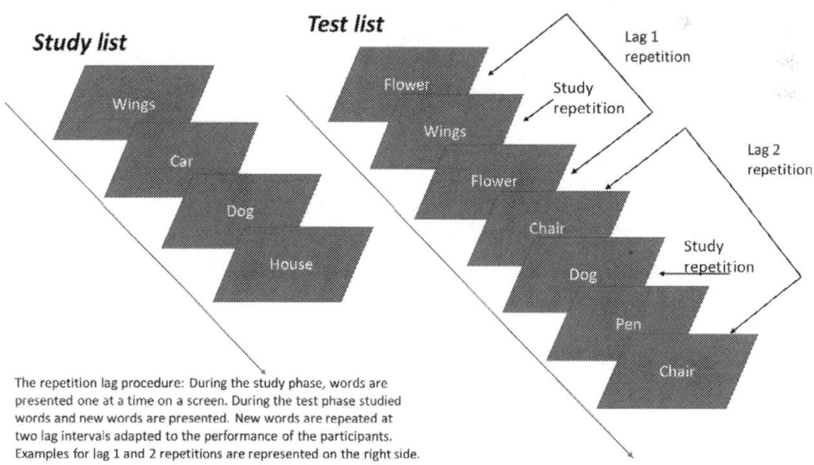

Fig. II.6.5 The repetition lag procedure (Stamenova et al., 2014).

presented words (which were not part of the study list) are repeated like the words from the study list. The distance (the number of words in between, or "lags") of these distractor repetitions is varied and adapted to the individual performance of the participants. In general, the participants start with a small number of words between the distractor repetitions, and the goal of the repetition lag procedure is to increase the number of intervening words to a maximal lag (with a remaining accuracy of correct rejection of about 80%). The problem for the participants is to distinguish if the repetition stems from the study list or from the earlier parts of the recognition list.

Such a distinction assumes that the participants consciously remember the context of presented words (e.g., through the question: Has this repeated word been a part of the study or the recollection list?). Thus, the repetition lag procedure focuses explicitly on the recollection performance and not on any retrieval at all (produced by familiarity or recollection, something which is not distinguished explicitly in other approaches for memory rehabilitation). This is not the only difference between the repetition lag procedure and other treatments of memory impairment. The participants in a repetition lag procedure training are not taught any type of mnemonic strategies, and they are not being educated how to organize information during encoding. They are simply instructed to do their best in discriminating between correct and false repetition. Therefore, the repetition lag procedure is a type of adaptive drill and practice training with a focus on contextual information to boost recollection instead of familiarity. Fig. II.6.6 shows how the repetition lag procedure can be integrated into the model by Atkinson and Shiffrin (1968). The repetition lag procedure acts on the ability to recover contextual information. It focuses on selective search in a retrieval buffer for features that are helpful in distinguishing between mere familiarity or familiarity and a specific encoding context.

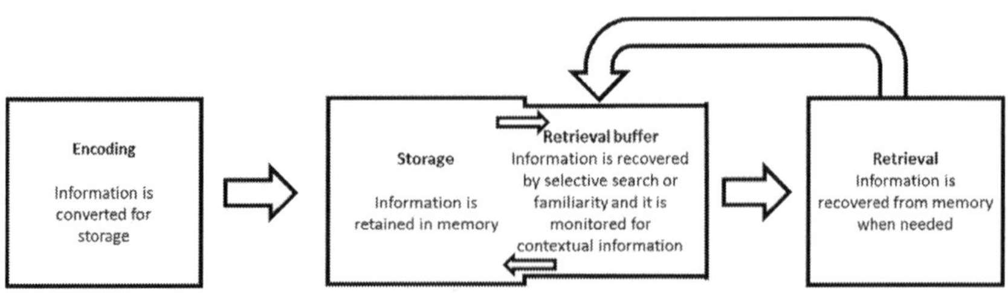

Fig. II.6.6 The augmented Atkinson and Shiffrin (1968) model and the putative role of the repetition lag procedure on training of memory performance.

There are several treatment studies using the repetition lag procedure which include some involving older adults, some in the study of neurodegenerative diseases, and some working with stroke patients. One of the first studies (Jennings et al., 2005) compared the results of a repetition lag procedure with a nonadaptive recognition practice group and a passive control group healthy subjects. The study was not a RCT, because the passive control group was added later to the design. Allocation concealment was not reported. The main outcome variable was performance in other neuropsychological tasks. This included a *n*-back task (which counted as a working memory task, but may also be solved by episodic memory), a self-order pointing task (which is also a type of working memory task but for visuospatial information), a source-monitoring task (a recognition task to discriminate items from two different lists and new items), the digit symbol substitution task, a reading span task, and the CVLT.

The repetition lag procedure was effective in the sense that the participants could discriminate between studied items and distractor repetitions with an increasing lag between irrelevant repetitions. In other words, the participants learned to distinguish between familiar (but false) items and presented (and correctly recollected) items with growing distances for relevant and irrelevant repetitions. This may indicate an improvement in their recollection performance. There were also significant interaction effects for the *n*-back task, the self-order pointing task, the source-monitoring task, and the digit symbol task. Post-hoc intragroup comparisons showed that only the repetition lag procedure group improved in these tasks. Such a statistical evaluation is incomplete in EBM terms, because no intergroup comparisons for the post assessment were performed. The main goal of an intervention is to show that the treated group is afterwards in a better state compared to a nontreated group, not that the treated group has just changed in its performance. Simply showing significant repeated ANOVA interaction effects may be driven by small losses and gains of performances at T2 combined with small differences in functioning at T1. Moreover, the training did not increase performance in the CVLT, but mainly in tests that rely on similar task structures as the repetition lag procedure itself. Therefore, no far-transfer effects were present.

In a recent study, Anderson, Ebert, Grady, and Jennings (2018) replicated the findings of Jennings et al. (2005) using a different version of the repetition lag procedure. This study also concerns recollection training in older adults and the items were presented either as written or spoken words. Therefore, recollection concerned only the meaning of the words, because the modality of presentation was changed. The performance of a group of younger healthy participants served as a benchmark (or clinical relevant outcome measure) for the impact of the training. The older adults trained with the repetition lag procedure for nine sessions. As a result, they improved in their recognition performance up to the level of the younger adults. A 3-month follow-up showed that the training gains remained stable for this period. On the other hand, as in the former study, there was no generalization to performance in other memory tests thought to rely on recollection. Moreover, the group who trained with the repetition lag procedure did not believe that their memory performance increased, something which was assessed by questionnaires. Although the achievement of the training sounds very impressive (equalizing the recollection performance of older and younger participants after just nine sessions), one should be aware that the younger participants were not trained. Therefore, we do not know their performance after nine sessions with the repetition lag procedure.

Two other studies using the repetition lag procedure focused on patient samples. The first study trained patients with Alzheimer's disease (Boller, Jennings, Dieudonne, Verny, & Ergis, 2012). The design of the study was very similar to that of Jennings et al. (2005). This study also involved three groups, but this time all patients were randomized to one of the groups. The specific algorithm for randomization was not explained, and the fact that gender was completely balanced for the three groups argues that randomization might have been not free of any systematic influence. Allocation concealment was not mentioned in the paper. No information about blinding of the assessor was given. The training duration encompassed 6 h.

Alzheimer's disease patients were able to increase the repetition lag for the distractors during the training (albeit on a very low level). The results in the *n*-back task, in recognition memory, in the delayed matching to sample task, and in the source-monitoring task showed a significant interaction for the groups. Group differences after the treatment were not tested. Intragroup comparisons showed that both active groups improved in the recognition part of the free and cued recall test. Only the group receiving repetition lag procedure training improved significantly in the *n*-back task, the delayed matching to sample task and the source monitoring task.

Stamenova et al. (2016) used the repetition lag procedure in a study involving stroke patients. Their results were compared to an age-matched control group. Thus, this study did not include blinding or randomization. Stroke patients improved as fast as the healthy control group. There were some indications for near and far transfer on other neuropsychological tasks.

Stamenova et al. (2016) concluded that recollection can be increased in stroke patients as well as in healthy controls, but they did not discuss whether a specific impairment in recollection is the main cause of the memory deficit in the patient group.

Summarizing the results of the repetition lag procedure, this method might be helpful in enhancing the performance of recollection relative to familiarity. Additionally, it might also show some generalization effects. To date, however, there have not been any RCTs of Level Ib that can show the effectiveness of this method.

Richter et al. (2018) used the repetition lag procedure in combination with WoMe to train patients with memory impairment resulting from acquired brain injury (mainly stroke patients, but also some patients with intracranial bleedings and TBI). They included patients with mild to moderate impairments as defined by the results of the CVLT. They used daily life memory performance as the primary outcome variable, which was defined by some of the tasks of the Rivermead Behavioral Memory Test and by some tasks adapted to the specific situation of the patients at the rehabilitation unit (where the training was provided). Thus, the primary outcome variable was everyday memory performance and not the transfer to other neuropsychological tests. Thirty-six patients were randomized to two different groups, with concealment ensured by opening envelopes (containing group assignment) only after the recruitment was completed. The mean age of the groups was about 51 years old, the average time since lesion was 31 days, and the memory impairments were typically mild to moderate (5th learning trial in CVLT: at maximum 10 words, short-term free recall: at maximum 8 words). Both groups received 9 h of treatment. The experimental group received computer-based WoMe training and the repetition lag procedure, while the active control group received the usual memory rehabilitation of the rehabilitation unit. After the treatment, the patients in the experimental group showed a significant increase in their everyday memory performance, i.e., there was a significant interaction in the ANOVA for repeated measurement and a group difference for the second assessment (Fig. II.6.7). Both groups improved significantly in word fluency and verbal learning. For working memory, the performance increase resulting from the repetition lag procedure/WoMe training was significantly larger than that of the active control group. Both groups also differed significantly at t2. Unfortunately, Richter et al. (2018) did not include a follow-up assessment (after 3 months or so) to analyze whether the improvement remained stable over time and if there were differences in job status between the groups as a result of the treatment.

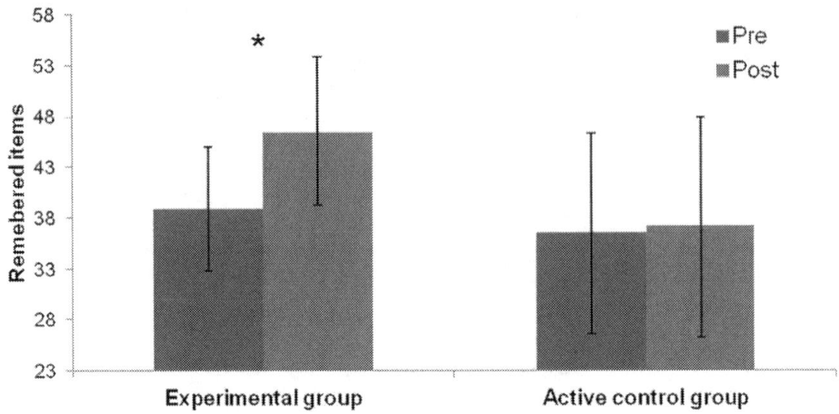

Fig. II.6.7 Improvement in everyday memory tasks (Richter et al., 2018).

Therefore, the repetition lag procedure/WoMe training of Richter et al. (2018) showed far transfer to everyday memory tasks and near transfer to working memory tasks in an RCT for patients with acquired brain damage. The latter finding fits the earlier studies with the repetition lag procedure in healthy participants showing near transfer to working memory (Anderson et al., 2018; Jennings et al., 2005). One may interpret the lack of a superior improvement in the CVLT also as in-line with the results of previous studies, which failed to show a generalization to performance in memory tests. However, in this respect it should be noted that the repetition lag procedure group improved *as much as* the standard group receiving memory training and that the difference between pre- and post-testing was significant for both groups. The important finding is that the repetition lag procedure combined with the WoMe training enhanced everyday memory performance. This would either mean that working memory played a role for the everyday memory tasks used in this study or that the CVLT as a test is too specific to assess the many different components of episodic memory.

Some direct comparison between the repetition lag procedure/WoMe training, the semantic structuring/WoMe training, and the semantic structuring/transfer appropriate processing training is possible because the studies used the same neuropsychological tests. Both treatments involving WoMe (Richter et al., 2015, 2018) did enhance verbal learning and recall only to the same extent as the standard memory treatment. This is still interesting, because in both studies the experimental group did not receive a specific verbal memory training at all. Both treatments led to improvement in ecologically important memory tasks (prospective memory or daily life memory tasks). On the other hand, the specific effect on verbal learning was large in the earlier semantic structuring/transfer appropriate processing training studies (Hildebrandt et al., 2006, 2011). Compared to the effect size of >1 in the CVLT for the latter training (Figs. II.4.10 and II.4.11), the effect size in the CVLT was smaller than 0.3 in the repetition lag procedure (Richter et al., 2018) and also in the WoMe training (Richter et al., 2015). Therefore, the impact on verbal learning and recall seems to be higher for the semantic structuring/transfer appropriate processing training. Unfortunately, these studies did not include ecologically more valid memory tasks except for the Hildebrandt et al. (2011, first study) study, which showed an improvement in the rating of the nurses.

7 MEMORY REHABILITATION FOCUSING ON RETRIEVAL TRAINING

7.1 Autobiographical Memory Training and SenseCam

Is it possible to treat memory impairments not by focusing on encoding or consolidation, but rather just on retrieval processes? At first glance, this question seems to be nonsensical, because one must first *learn* a piece of information before it is possible to *retrieve* it from memory. Therefore, almost all memory rehabilitation approaches have concentrated on improving encoding instead of retrieval.

On second thoughts, however, a memory rehabilitation program that focuses exclusively (or at least primarily) on retrieval is neither practically impossible nor theoretically incorrect. Consider one of the generally accepted neuropsychological interventions for patients with Alzheimer's disease: the "Selbsterhaltungstherapie" (Romero & Wenz, 2002). This intervention uses existing biographical information of the patients to help them recall former life events and thus to sustain a kind of autobiographic identity. Reactivation of former memories is also a core method of music therapy for demented patients, and this method seems to be a powerful intervention for improving quality of life in this patient group. Although the main goal of "Selbsterhaltungstherapie" in Alzheimer's patients is to stabilize autobiographic information and personal identity, it also offers the possibility to train retrieval processes without encoding or new learning. After all, a similar approach as that of the "Selbsterhaltungstherapie" is frequently used in acute patients with severe TBI or hypoxia. In acute patients with severe brain damage and dense anterograde amnesia such retrospective recollection practice (e.g., using photographs of family members, etc.) is often the only possibility to motivate the patients for memory training. Until now, nobody seems to have evaluated the effect of such a treatment on the global recovery of memory performance. There is yet another approach to use guided retrieval attempts for the remediation of (autobiographical) memory performance: In the last few years, several researchers used "SenseCams" mounted on the chest or shoulder of memory-impaired patients. The cameras collected pictures throughout the day with the purpose of cueing retrieval of the day's events (Dubourg, Silva, Fitamen, Moulin, & Souchay, 2016; Silva, Pinho, Macedo, & Moulin, 2018). Several SCEDs show that this kind of cueing may help to recover information from memory and to improve autobiographical identity in amnesic patients. However, as with the "Selbsterhaltungstherapie", in patient studies involving "SenseCam" treatment the focus was primarily how to enhance memory for specific episodes experienced throughout the day. There was only one trial to look for a possible generalization to memory performance outside the direct application of the SenseCam (see below).

Silva et al. (2018) argued in their review of the papers published on SenseCams that in order for them to qualify as memory rehabilitation instruments, their use should also lead to an improvement in the recall of features and events that are *not* represented in the pictures taken over the day. In other words, the cueing effect of the pictures for retrieval should lead to an improvement of memory retrieval in general. In the previous study from 2013, Silva et al. showed that when healthy controls review pictures taken by a SenseCam, it leads to improved memory performance in classical neuropsychological tests like the CVLT. They therefore conceptually compare treatment with SenseCams with the treatment of aphasic patients or neglect patients. In these cases, the improvement of the impaired function is initiated by helping the patients with systematic cueing hierarchies and trying to fade out these cues at an increasing rate over time (Fig. II.7.1).

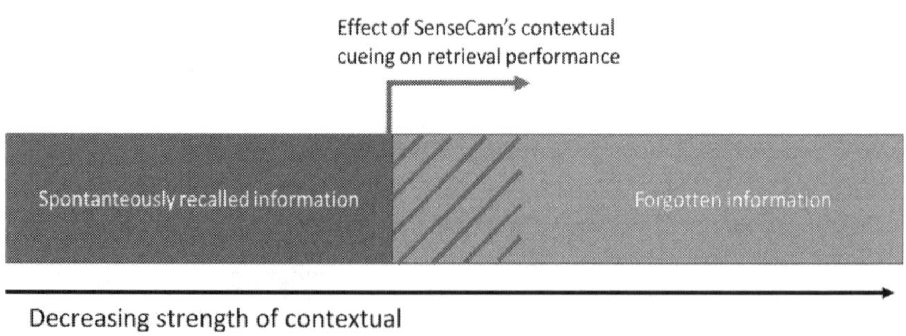

Fig. II.7.1 Possible cueing effects of wearing a SenseCam on memory performance (Silva et al., 2018).

In a recent RCT, Silva, Pinho, Macedo, and Moulin (2017) tested the effects of SenseCam on other memory performances. In this study, 60 patients with mild Alzheimer's dementia were randomized into three groups. Participants in the first group were provided with the SenseCam to capture images from their daily life for the whole period of the intervention (6 weeks), and were reviewed two times a week (for a total number of 11 reviews, each lasting 1 h). The second group received training of attention, working memory, autobiographical and episodic memory, semantic memory, and implicit memory with the same amount of treatment time as the SenseCam group. In the third group, participants were asked to write down their daily activities in a personal journal and to then read it to the experimenter during the meetings. The SenseCam group showed a superior autobiographical memory performance directly after the training and also at a 6-month follow-up. This group and the second active training group achieved improvements in episodic and semantic memory. The SenseCam group displayed improvements in verbal fluency, while the active training group showed enhanced performance in working memory. The third group, whose members did not receive active training (but instead were instructed to reread their diaries), declined in many cognitive functions during the 8-month test period. This decline is expected as a normal disease progression of patients suffering from a neurodegenerative disease. On the one hand, the results of the two other groups are impressive for patients with Alzheimer's disease, and they show that a cognitive treatment may improve memory functions. Still, a few key methodological problems are present in this study. There was no blinding of the assessor, and this is especially a problem for the results that are based on cueing (as for example, the autobiographical memory test). When the assessor is not blinded for the treatment status, it might lead to a difference in behavior during cueing. Naturally, that may have influenced the results. Furthermore, no information is given about concealment, and it is not clear whether randomization of individual patients to groups took place before or after the neuropsychological assessment.

Coming back to the review of Silva et al. (2018), there is another important aspect that should be mentioned. When unimpaired, healthy controls were confronted with pictures taken during their own day or during the day of another person, they were only able to classify the pictures from their own day with 80% accuracy. In some other studies, it became obvious that patients tend to rely on semantic memory and familiarity in trying to remember the context of the information. This leads back to the dual trace model of memory (recollection and familiarity) discussed above. Cueing with first-person perspective pictures may not only be powerful in enhancing recollection, but also in enhancing familiarity. This may therefore induce an increased rate of memory errors. This might be a challenge when using wearable cameras in heterogeneous groups of patients with memory impairments, and it must be analyzed in future studies.

7.2 Spaced Retrieval

The assumption that retrieval of information might be a cognitive function that is impaired and should be treated separately can also be endorsed on a conceptual level. There are many memory-impaired patients that show intact recognition performance in neuropsychological tests, but impaired free recall (compare Silva et al. (2017) for a similar argument to introduce SenseCams as reasonable apparatus for memory rehabilitation). The problem of these patients seems to be accessing stored information and not so much the encoding and consolidation of that information. Actually, there is an important memory technique that focuses on retrieval (and not on encoding): The spaced retrieval method. This method relies on the observation that recall performance is optimal if the already-learned information is accessed again and again with an expanding lag between the retrieval attempts. Increasing the intervals between the attempts leads to a higher recall performance than just retrieving it several times without pausing. Karpicke and Roediger (2008) published an influential paper that compared the effect of restudying (encoding) or recalling (retrieval) information on long-term recall. In their experiment, students learned from vocabulary lists of word pairs. The performance in delayed recall for the translation of the words from the new language into the native language was assessed 1 week later. The two critical variables (encoding and retrieval) were varied by dropping words that had been correctly learned either from the future presentation of the word lists or from future retrieval attempts. In other terms: If a student retrieved a translation correctly, this word pair was either not shown any more during the next learning trial (but it was retested), or the pair was shown again during the next learning trials (but not retested). The question was which method would be superior for long-term retention. The results were quite clear-cut: Dropping the word pairs from relearning but testing them again and again during the learning phase outperformed restudying. This would mean that the retrieval process plays a major role in consolidation and the effective use of the spaced retrieval method mirrors exactly that finding for and in clinical practice.

While the spaced retrieval technique plays a major role in memory rehabilitation as a technical trick to teach severely memory-impaired patients specific information (like the "Selbsterhaltungstherapie" and in some studies the use of a SenseCam), it is not used for treating memory deficits per se. The therapists usually implement the procedure in order to achieve *specific goals* of the patient, and this use is not similar to improving their overall memory performance. Han et al. (2014, 2017) and Lee et al. (2009) performed studies to analyze whether the application of this procedure would also transfer to memory performance measured with neuropsychological tests after the end of the training. Patients in these studies suffered from Alzheimer's disease or mild cognitive impairment. Two of the studies used a tablet and computerized training software (Han et al., 2014, 2017), while the other (Lee et al., 2009) was a group program, i.e., involving the presence of a therapist. Only the Han et al. (2017) study was single-blinded, randomized, and controlled. The other two studies had no control group and therefore the results of these studies are only partly informative. Han et al. (2017) included 43 patients, divided into 23 verum patients and 20 patients for the control group, but the design was a cross-over for intervention and waiting list control. The blinding concerned only the assessor. Patients were not blinded, because the control group did not receive specific memory training. As randomization was performed by a computerized algorithm, concealment seemed to be guaranteed, although this was not described in detail. The outcome measures were wordlist learning, wordlist free recall and wordlist recognition of the CERAD-NAB (two parallelized versions). Other measures concerned the Subjective Memory Complaint Questionnaire and the Geriatric Depression Scale. A follow-up assessment was not performed. Therapy intensity was low: 30 min per session, twice per week, over a 4-week intervention period. The increase of the retrieval lag followed a fixed sequence, but the difficulty of the memory tasks was automatically adapted to the performance of the patients: In each session, the retrieval interval was doubled from 0.75 to 12 min (i.e., 0.75, 1.5, 3, 6, and 12 min). If the participant succeeded in recalling a given set of words within 12 min in two consecutive sessions, the number of words to be memorized within a session was automatically increased in the following session from one to five words. The two studies that concerned mild cognitive impairment patients showed improvement in word list learning of the CERAD-NAB (Fig. II.7.2, which shows the results of Han et al., 2017. This study was a cross-over study with a verum and a waiting list control group. Therefore, the dotted line concerns performance of the waiting list control before and after the treatment.), but

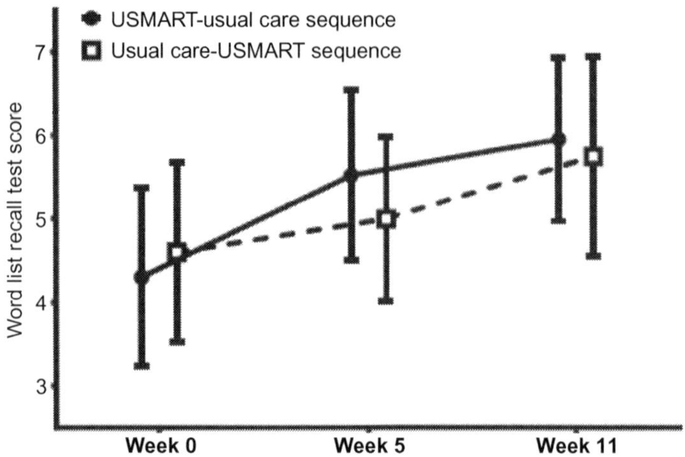

Fig. II.7.2 Results of Han et al. (2017). WLRT scores during the trial according to the treatment sequence. Bar: 95% confidence interval. Week 0 the beginning (baseline), Week 5 the end of the first treatment period, Week 11 the end of the second treatment period, USMART Ubiquitous Spaced Retrieval-based Memory advancement and Rehabilitation Training.

not in the Mini Mental Status Examination, the Subjective Memory Complaint Questionnaire, and the Geriatric Depression Scale. Improvement in word list learning correlated with the number of treatment sessions (Han et al., 2014). The study on patients with Alzheimer's disease (Han et al., 2017) documented short-term clinical improvement of word recall, but no effect on neuropsychological test results. Effectiveness, therefore, was associated with the severity of cognitive impairment (only mildly impaired patients profited from the procedure) and with the intensity of the treatment (it took many sessions to yield improvements). Moreover, partly due to the type of post-interventional testing, only near-transfer effects could be documented (no extensive neuropsychological test battery was used; recognition performance of the patients was almost perfect, thus improvements impossible; increase of free-recall performance was present only in absolute terms, but not as a significant difference). In general, for the effect to be expected from such a low number of interventional sessions (eight sessions of 30 min each, distributed over 4 weeks) the study was massively underpowered.

Fig. II.7.3 shows a potential mechanism for how spaced retrieval might work. Because the impact of repeated access to information stored in the episodic buffer will depend on the dynamics of decay, the question as to when and how often a retrieval attempt takes place may play a central role.

Han et al. (2014, 2017) and Lee et al. (2009) tried to automate the spaced retrieval lag by systematically increasing the time between the retrieval attempts and using the number of words to be remembered as a variation of the task difficulty. This might be the suboptimal strategy for memory rehabilitation using the spaced retrieval technique. The repetition lag procedure, for example, tries to adapt to the patients by increasing the lags. Instead of a fixed sequence, this procedure employs a flexible algorithm that is related to the performance of the patients. The specific conceptualization of the "space" between the retrieval attempts may be of major importance for using this strategy as a memory training approach. Such an adaptive spacing procedure has not been realized in the Han et al. (2014, 2017) study and it would be interesting to see whether it would increase the effects of the program.

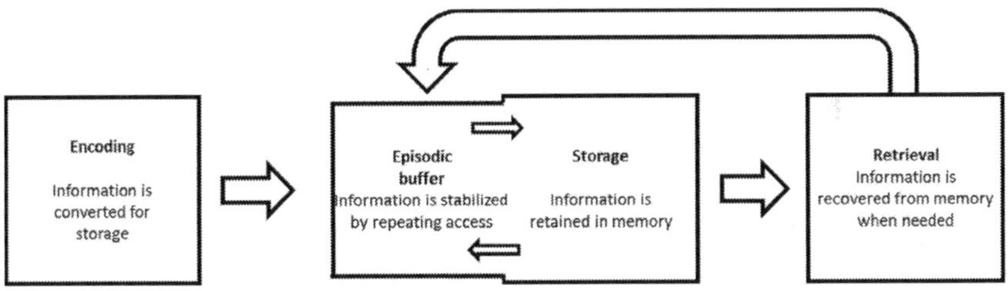

Fig. II.7.3 The augmented Atkinson and Shiffrin (1968) model and the putative role of spaced retrieval for memory performance.

The essential role of a considerable residual retrieval performance for the efficacy of retrieval practice and spaced retrieval has been shown by Maddox, Balota, Coane, and Duchek (2011), who compared the memory performance of healthy young and healthy old participants for different lags in a spaced retrieval experiment. They found that a very essential aspect for the efficacy of spaced retrieval is the length of the *first* lag: In older adults, the presentation of two intervening items between first retrieval attempts produces dramatic rates of forgetting compared to one intervening item. In contrast, younger adults can maintain performance for up to five intervening events in comparable conditions. Thus, the position of the first spaced retrieval attempt must be adapted to the individual memory performance of the patient, otherwise the spacing of the next attempts no longer plays a role. Maddox and Balota (2012) allowed younger and older subjects to choose the frequency and timing of retrieval attempts during a distracting secondary task and compared the result of the freely chosen spaced retrieval lags with a sequence that was defined by the experimenter. Three results emerged from that study: (1) Both groups showed a better memory performance following their individual arrangement of retrieval attempts. This means that both groups were able to monitor their memory performance during the secondary distracting task and to decide at which timepoint a retrieval attempt was necessary to prevent forgetting learned information. (2) Both groups decided to have an early retrieval attempt before starting to expand the lag differences. (3) Performance in the distracting secondary tasks was not reduced by freely choosing an individual sequence of spaced retrieval attempts. The results of this second study underline the importance of an early attempt to stabilize the memory trace. Moreover, the study showed that healthy participants have a "feeling" for the optimal time lag between retrieval attempts. This seems to argue that there is an automatic signaling or monitoring process for the status of a memory trace. It would be very interesting to know whether this automatic monitoring process is intact in at least some patients with memory impairments, as well as whether (1) they can be taught to use it for the individual adaptation of the method of spaced retrieval and (2) if such an automatic monitoring process was lost as a consequence of brain disease, if it is possible to retrain it.

Unfortunately, until now, the goals of the studies on spaced retrieval were either to teach the patients information or to improve their memory function, but there have been no major attempts to teach the method as a technique for the patients themselves. To our knowledge, there is only one small study that aimed to accomplish this (in the studies of Hildebrandt et al. (2011, 2006) the patients were instructed to individually determine how long the lag between retrieval attempts should be in order to not forget the information, but spaced retrieval was only a minor aspect of the treatment. Therefore, it is uncertain as to exactly what role this method played in achieving the positive results.). Schacter, Rich, and Stampp (1985) trained four patients with organic memory disorders using the spaced retrieval technique to see whether it is effective in enhancing the number of reproduced words (such as names, hobbies, and social status belonging to presented faces). They also wanted to see if the patients could learn to learn, i.e., to use the technique without any external cueing. Two of the patients in this study were severely memory impaired, while the other two patients had mild to moderate impairment. Schacter et al. (1985) used a systematic cueing system after the presentation of the faces for retrieval attempts but also tried to fade out these cues and to establish a spontaneous use of the spaced retrieval technique by the patients themselves. Initially, they started with the presentation of the faces as cues for retrieval attempts and prompts were given by the experimenter. Next, the experimenter only prompted the patients verbally, finally reducing even these prompts. As the patients had to solve distraction tasks between the testing sessions, Schacter et al. (1985) could therefore score the number of spontaneous retrieval attempts. These were defined as interruptions in solving the distraction tasks. The spaced retrieval method led to an increase in remembered information concerning faces. The impact was dependent on the lag between the presentation and the first retrieval attempt. Long spacing had no impact on the performance. The two severely impaired patients did not learn to use the spaced retrieval spontaneously (at least not in the treatment time given to learn it), but the two mildly to moderately impaired patients were able to use it without prompting anymore by the experimenter.

Given the high impact of spaced retrieval on memory performance, the study of Schacter et al. (1985) conveys important findings for the rehabilitation of memory-impaired patients. Unfortunately, it is only a series of single case trials and we do not know if it is possible to teach the strategy of spaced retrieval to a noteworthy portion of memory-impaired patients. If this would be possible, that would be an important step for the internal compensation of memory deficits in patients with mild to moderate impairment.

There is one other important aspect that should be mentioned: neither Han et al. (2014, 2017) nor Lee et al. (2009) found a difference in the discrimination index of the CERAD-NAB after the retrieval training. Thus, one may conclude that this type of retrieval training does not exert aversive side effects such as an increased role of familiarity (instead of recollection) for retrieval performance. Bishara and Jacoby (2008) analyzed the effect of retrieval practice of one list on retrieval practice of another list. The first list was read by the participants themselves, while the second was presented to them verbally by the experimenter. During the final test, the participants were supposed to only retrieve the words from the second presentation, i.e., ignore the words from the first list they had read themselves. Half of the participants were young adults; the other half consisted of older adults. Retrieval practice of the first list increased intrusions in the test for older adults. Results suggest that because of a deficit in recollection, older adults were less able to oppose the automatic influence of spaced retrieval. This would argue that spaced retrieval should only be used in patients suffering from a retrieval ("access") deficit and not in patients suffering from a consolidation–recollection deficit.

8 SPECIFIC GUIDELINES FOR REHABILITATION OF MILDLY TO MODERATELY MEMORY-IMPAIRED PATIENTS

We begin this section with a diagram of the German guideline for diagnosing and treating memory disorders (Fig. II.8.1).

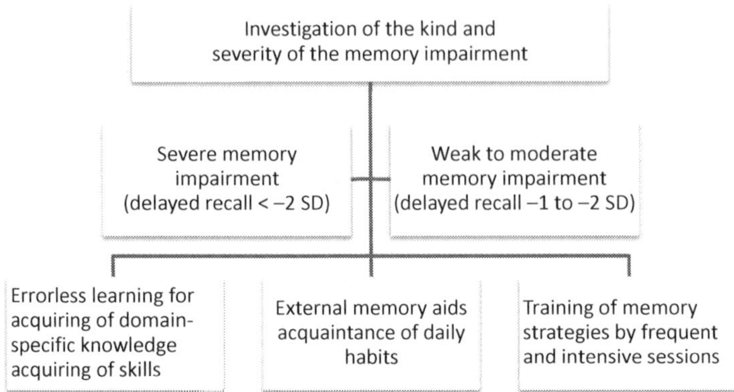

Fig. II.8.1 Flowchart for treating memory impairments from the German Guideline *Diagnostik und Therapie von Gedächtnisstörungen.*

We now have the possibility to slightly enhance this flowchart for clinical practice. Apart from the indication for different interventions according to the severity of the memory impairment, how many treatment sessions are possible and how many of them should be face-to-face are important questions to consider in clinical settings. Table II.8.1 gives an expanded overview on rehabilitation techniques and evidence levels according to the severity of the impairment. Some of the mentioned interventions concern the treatment of severely impaired patients. We will discuss the evidence for these kinds of interventions in the next section.

Table II.8.2 is divided into two major sections: one for patients in a post-acute and one for patients in a chronic state. Only for patients in a post-acute state is there still a chance for remission and therefore one will focus on direct training of memory, especially if the impairment is weak. It should be mentioned that the book by Wilson (2009) on memory rehabilitation concerns patients in the chronic phase suffering from moderate to severe memory impairments and that the reader will find in it a lot of helpful information how to treat such patients using a cognitive behavioral approach. The second subdivision concerns the severity of the impairment. In patients with mild memory impairment, teaching a compensatory mnemonic is evidence

Table II.8.1 Evaluation chart for the evidence of the studies on memory rehabilitation according to severity of impairment and treatment intensity

Impairment	**Severe**		**Moderate to mild**				**Mild**			
Estimated time for treatment	**High**	**Low**	**High**		**Low**		**High**		**Low**	
Face to face	High	High	High	Low	High	Low	High	Low	High	Low
	EL/Memory aids	EL/Specific tasks	SS/TAP	WoMe/TAP	SS/TAP	RLP/WoMe	Visual mnemonics	WoMe/TAP	RPL/WoMe	RLP/WoMe
Outcome quality level										
Effect size of treatment										
Estimated EBM Level										
Practice standards										
Cochrane 1										
Cochrane 2										
Cochrane 3										
Cochrane 4										
Cochrane 5										
Cochrane 6–8										

Table II.8.2 Impairment severity, evidence-based treatment strategy, and treated memory process

Impairment	Treatment strategy	Evidence level	Problems	Treated process
Post-acute and mild	Working memory training and repetition lag procedure	Ib (Richter et al., 2018)	No follow-up small group size	Encoding
Post-acute and moderate	Semantic structuring and transfer appropriate processing; Working memory training and Semantic structuring	Ib (Hildebrandt et al., 2011, first study); Ib (Richter et al., 2015)	No follow-up; small group size No follow-up, clinical relevance unclear	Specific subroutines of encoding and retrieval Encoding
Acute and severe	Improving orientation Source memory training in the case of confabulations	IIa (Langhorn, Holdgaard, Worning, Sorensen, & Pedersen, 2015) IIa (Trivino, Rodenas, Lupianez, & Arnedo, 2017)	No randomization No placebo group, small group size Small group size, inclusion criteria rather unspecific	Knowledge about data and location Source memory
Chronic and mild	Visualization mnemonics; modified Story Memory Technique; Training of global encoding strategies	Ib (Kaschel et al., 2002); Ib (Chiaravalloti et al., 2013, 2016) IIa (Berg et al., 1991)	Very small group; Follow-up negative; Clinical relevance unclear	Additional mnemonic strategy Unspecific processes involved in encoding
Chronic and moderate	Semantic structuring and transfer appropriate processing	IIa (Hildebrandt et al., 2006)	No follow-up, clinical relevance only in form of text reproduction	Specific subroutines of encoding and retrieval
Chronic and severe	Paging system as external memory aid Errorless learning	IIa (Wilson, Emslie, Quirk, & Evans, 2001; Wilson, Emslie, Quirk, Evans, & Watson, 2005) III (Glisky & Schacter, 1989)	No placebo group, restricted outcome measure Single case study	External compensation and procedural learning

based as show by several Level Ib studies. Patients with moderate memory impairment should be treated with easily learnt memory strategies boosted by bottom-up training of semantic structuring and forcing them to stick to a common organization during encoding and retrieval. The group with severe memory impairment and being in an acute or post-acute state is a special group. These patients suffer not (only) from amnesia but often from a delirium-like confusional state. Helping the patients to regain orientation is the primary goal of intervention then. Because confabulations are frequently accompanied by an increased risk of potential self-hurting behavior, a second aim is the reduction of such spontaneous false beliefs. Scientific studies on treating confabulations are almost absent.

CHAPTER III

Compensation of Memory Disorders Through External Memory Aids

1. SPONTANEOUS USE OF MEMORY AIDS

Compensating for memory problems with external aids is common in memory-impaired patients. Wilson and Kapur (2009) showed that these patients tend to use external memory aids spontaneously, often and similarly to healthy controls. About 72% of the examined memory-impaired patients used wall calendars, 64% used paper and pencil notebooks, and 63% used task lists. Although this was an older study (and smartphone technology was not as ubiquitous as it is today), 7.5% of the patients used electronic memory aids. A recent study by the same group revealed that the most often used external memory aids are asking someone else for a reminder (78%) or using diaries (77%), lists (78%), and calendars (79%). This study also revealed that 38% also used mobile phone reminders and timers (38%). The percentage of electronic technology use correlated negatively with age. The inclusion criteria in this study were (1) suffering from an acquired brain damage, and (2) self-reported memory impairments. de Joode, van Heugten, Verhey, and van Boxtel (2010) provide similar figures for the use of electronic memory aids in a systematic review on this topic.

Recently, Wong et al. analyzed the spontaneous use of smartphones in stroke patients and patients with traumatic brain injury (Wong, Sinclair, Seabrook, McKay, & Ponsford, 2017; Wong, Wang, Stolwyk, & Ponsford, 2017). Usage did not differ in the group of younger patients with traumatic brain injury (TBI) compared to an age-matched control group. In the case of stroke patients, fewer patients used a smartphone than in the age-matched control group. Age was an independent factor for smartphone use regardless of disease status. For both diseases, patients more often answered that they used the smartphone as a memory aid. The patients that rated themselves as more impaired more often reported difficulty in using a smartphone.

Cognitive Rehabilitation of Memory
https://doi.org/10.1016/B978-0-12-816981-0.00003-5
© 2019 Elsevier Inc.
All rights reserved.

Collerton, Forster, and Packham (2014) conducted research on the effectiveness of memory aids for older people living in supported accommodations. This study is interesting because the group suffered from low memory impairment, low additional cognitive impairment, and no emotional complaints due to brain damage. The spontaneous use of memory aids in this group of older people was almost the same as in memory-impaired patients: 74% used mnemonics, 64% used calendars, and 52% used notes and lists. After the therapeutic introduction of *electronic* memory aids, their usage led to an improvement in concentration, a reduction of misplaced items, improved memory of names and faces, and increased attendance of appointments. Nevertheless, the success of the training program was low, because the participants were unwilling to use additional electronic memory aids in everyday life. Their interest was more focused on using a computer but not on its use as a tailored memory aid. This shows that treatments concentrating on memory aids, i.e., external compensation for individual goals, still struggle with the problem of transferring to everyday life. This is a very real problem, despite that it theoretically should only be present in internal compensation or restitutive trainings.

2. GROUP STUDIES ON TEACHING EXTERNAL AIDS TO MEMORY-IMPAIRED PATIENTS

2.1 Teaching Memory-Impaired Patients to Use Electronic Memory Aids Reactively

Using external memory aids for severely memory-impaired patients started with the introduction of calendars and "memory books", i.e., a more systematic use of noting systems also used by nonmemory-impaired patients. We will discuss the prospects but also the pitfalls of such an external memory aid system for a patient with very severe memory impairment in the next chapter. Wilson, Evans, Emslie, and Malinek (1997) introduced an early electronic memory aid system called "NeuroPage" to a group of memory-impaired patients. NeuroPage is a pager system, which is updated via the internet (or its precursors) and it works by sending one message to the recipient (Fig. III.2.1).

The interaction with a pager through NeuroPage is very limited: One may press a button for OK or to receive a new reminder in a few minutes. The messages can be uploaded to an internet calendar and are automatically sent to the recipient at predetermined times. Access to the list of reminders is gained only through a computer, which is modified by a significant other or

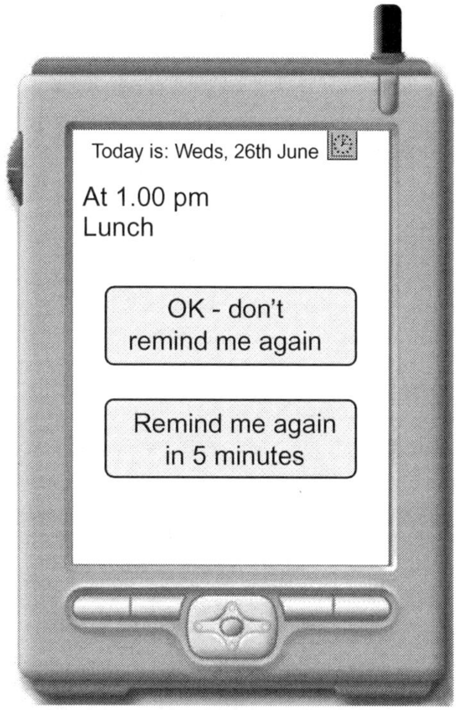

Fig. III.2.1 NeuroPage showing a message (Morrison, Szymkowiak, & Gregor, 2004).

by a service collecting the reminders for several patients using the system. In a first study, Wilson et al. (1997) analyzed the impact of NeuroPage on 15 TBI patients in a single-case experimental design (SCED) cross-over design, in which one half of the group served as a waiting list extended baseline control. In the first step, one or a small number of important goals were identified together with the patients and their relatives. These goals were to be achieved through the introduction and use of NeuroPage. (Example goals include: eating my lunch, taking my drug(s), switching off the lights, etc.). In the next step, the participants received goal attainment diaries, in which they and their relatives logged how often the defined goal was reached without the use of NeuroPage. Next, the pagers were handed out to half of the patients and their usage was explained. The pagers were then used by this group of patients for 7 weeks. After these 7 weeks, the waiting list patients received the pager. Using the pager increased the probability of achieving the predetermined goal (or goals) from 37% to 85%. Interestingly, the group that started with the pager and stopped using it after the crossover of the conditions still attained the defined goal in 74% of the cases. In another publication (Teasdale et al., 2009) it was shown that the use of the pager decreased the Caregiver Strain Index by about one point (out of a maximum of 13 points, which means a decrease of 8%). In several different larger group studies of patients with different etiologies, Wilson, Emslie, Quirk, and Evans (2001), Wilson, Emslie, Quirk, Evans, and Watson (2005), and Emslie, Wilson, Quirk, Evans, and Watson (2007) used the design of a randomized controlled trial to show the efficacy of NeuroPage in assisting memory-impaired and executive-function-impaired patients (Fish, Manly, Emslie, Evans, & Wilson, 2008). On average, eight messages were sent to the patients via NeuroPage per day. The evaluation of the efficacy of this system was conducted as described above. The results were always similar, with one exception: Somewhat older patients with a vascular cause of their impairments decreased in their performance after they stopped using NeuroPage.

What is the evidence level of this series of studies? Focusing on Wilson et al. (2001), the report states that the study was randomized. However, there was no concealment, as after patients were allocated to groups some of them were switched due to knowledge about their actual situation and their risk of dropping out. Moreover, the study was not blinded: The patients and their significant others always were aware whether they were receiving the treatment (used the pager or not). Because the evaluation was done by the patients themselves (with help of their significant others), the nonblinding of the studies may have influenced the results. Waiting for an intervention (as opposed to receiving it) might have an impact on judging whether one was successful in achieving a goal which was set in that study. In other words, patients who have not received the pager (that they already know is supposed to assist them in reaching their goal) may simply assume that they have not made any (or very little) progress towards their goal. In the Wilson et al. (2001) study, NeuroPage increased the success rate (achieving the defined goals) from 48% to 75%. This is somewhat less than in the first case series study (Wilson et al., 1997) and one also has to ask whether this is a clinically relevant difference: While it is better to take an important drug meant for daily use 75% of the time as opposed to only half of the time, when helping patients to become independent the benchmark is almost 100%. Therefore, the clinical relevance remains unclear. The same holds true for the reduction of the Caregiver Strain Index in the Teasdale et al. (2009) study. A reduction of less than 10% needs an external validation for verifying its clinical importance. Nevertheless, it should be noted that the patients in this study were severely memory-impaired as displayed by a screening score of less than 5 in the Rivermead Behavioral Memory Test. Therefore, NeuroPage may be useful for helping severely memory-impaired patients. The evidence level for this statement is Level IIa.

2.2 Teaching Independent Use of Electronic Reminders to Memory-Impaired Patients

Dewar, Kapur, and Kopelman (2016) investigated the effects of a systematic training of the use of compensatory memory aids on everyday memory function. Patients ($n = 128$) suffered from progressive or nonprogressive neurological diseases and were randomized to an intervention group or a control group. There were three training sessions to teach the use of the memory aid, lasting about 1.5–2 h each. In these sessions, specific important goals and the possibility of the memory aids to help in achieving these goals were identified and matched. Goal achievement was measured both after the treatment sessions and at a follow-up appointment 12 weeks later. The waiting-group control patients received no treatment initially, but they were aware that they would receive treatment after 18 weeks. The critical outcome measure was a goal attainment diary. The patients were reminded to fill out this diary with help of significant others, who were also aware of the patient's treatment status.

There were no significant differences between the intervention group and the control group in neuropsychological testing. The ratio between nonprogressive versus progressive disease in the intervention group was 42:17 (72% nonprogressive patients) while the control group had a ratio of 9:12 (which equals 43% nonprogressive patients). Only the groups with nonprogressive neurological disease showed any benefit, and the benefit was only noted at the follow-up. The different frequency of progressive and nonprogressive patients in the two groups might have influenced this result. Intervention did not lead to a consistent transfer effect on a problem-solving inventory questionnaire. Nonprogressive and progressive patients differed in prospective memory performance. Dewar et al. (2016) concluded that teaching patients a memory aid may be beneficial for memory-impaired patients.

The study of Dewar et al. (2016) shows several methodological problems which hinder any straightforward interpretation of the results. First, the critical measure that was used was completely subjective (entries in the goal attainment diary) and the patients and the significant others helping the patients were not blinded towards the treatment. On this basis, it is surprising that the results are only modest. Second, no transfer effects to other cognitive functions or everyday performance were found. Third, the subdivision of the patients into progressive and nonprogressive groups was a post hoc move, and therefore the originally defined outcome parameter was not reached. Fourth, more than one-third of the recruited patients dropped out, and there was a considerable difference between the number of drop-outs in the treatment group (around one-third) and the control group (around one-half). A last observation of carried forward analysis is not present in this study. On a conceptual level, it is surprising that memory-impaired patients could learn to use a memory aid in only three sessions distributed over 6 weeks. In this respect, it is of interest that the nonprogressive group, which seemed to profit from the treatment, showed less memory impairment than the progressive group. Therefore, the main difference between patients who showed some success and those who did not may not be progressive versus nonprogressive, but instead more versus less memory impairment.

de Joode, van Heugten, Verhey, and van Boxtel (2013) recruited 34 patients and randomized the patients to two groups: one using personal digital assistants, the other a paper and pencil memory aid. Most of the patients were in a chronic state and their mean age was about 40 years. Memory impairment was mild to moderate: The average delayed recall in a Dutch version of the auditory verbal learning test was about seven words and less than 50% showed a deficit in delayed recall. Outcome measures were taken before the training, after 8 h of training, after 16 h of training, and finally at a 5-month follow-up. The Planning and Execution Assistant and Trainer (PEAT) software was used for the experimental group. This software offers four main modules: cue cards, diary, notes section, and names section. There is some automatic control for double booking and for consistency of entry placement related to the expected time it takes to fulfill the tasks. Outcome measures consisted of achievement scales as scored by the patients. Randomization and concealment procedures are discussed in the paper. Twenty-three patients entered the intervention group and 17 were assigned to the control group. While this is an unusual misbalance, it might be a consequence of the separate randomization of the different contributing rehabilitation centers. More than 50% of the patients did not return for the follow-up. Directly after the treatment, questionnaires of 10 patients from the control group and 19 from the intervention group were analyzed, which suggests a risk of bias in the analysis (due to the different numbers of drop-outs). Both groups did not differ in goal attainment and other measures of anxiety, depression, or quality of life. Goal attainment increased in both groups from below 20% to about 60% after the treatment sessions. As a result, this first randomized controlled trial on teaching patients an electronic memory aid showed no advantage compared to paper-and-pencil compensation techniques. Furthermore, the goal attainment in this study was less than two-thirds. In cases where achieving the goal is of central importance (such as taking drugs, etc.) this result seems to indicate only limited success.

McDonald et al. (2011) tried to use Google Calendar as a memory aid for neurological patients. A randomized crossover design was performed in which Google Calendar versus a standard diary were used as interventions. The outcome measure was again the goal attainment scale scored by the patients. The use of Google Calendar led to improved goal attainment (from 60% to 80%), while just keeping a diary had no impact on goal attainment. The impact of Google Calendar differed according to the degree of impairment: Only the less-impaired patients benefited from the treatment. Memory impairment varied strongly between the recruited patients: 5/12 of the patients showed no memory impairment at all, 2/12 showed a mild impairment, and 5/12 a moderate impairment. Only one of the patients was impaired in the Wechsler Memory Scale Working Memory Index and only one in executive function. This means that the overall cognitive impairment of the group was low.

Although these figures seem to indicate that the external compensation of memory impairment is a "natural" way to help memory-impaired patients cope with their disability, there are other studies that argue for considerable limitations when focusing on such an intervention. For example, Del Felice et al. (2017) were unable to find convincing evidence that external memory aids are helpful for patients with temporal lobe epilepsy. A recent review on memory aids in demented patients (King & Dwan, 2017) states that most studies still focus on the usability of the aids and not on at-home outcomes or improvements in daily life. For patients with acquired brain damage, a similar review from Mahan, Rous, and Adlam (2017) argued that if the patients suffer from prospective memory impairment, external memory aids may be helpful. The Cochrane review of Linden et al. (2016) argued for low-level evidence in patients with traumatic brain injury. According to this review, the effect of the electronic aid shown in most studies may be a side effect of the involvement of the therapists and of the electronic device. Thus, the specific impact of memory aids and specifically electronic memory aids in patients with memory disorders is still not completely determined. At face value, they should be helpful in regaining independence, but the empirical evidence is still somewhat uncertain. The evidence level for teaching independent use of external electronic memory aids in patients with mild to moderate memory impairment seems to be Level IIa/IIb.

2.3 Are Compensatory Treatments the Only and Best Strategy to Achieve Compensation for Memory Impairment?

We may wonder if training memory-impaired patients to use memory aids is really the best method to ensure that the patients consistently employ a strategy to compensate for their impairment. Although this question might sound somewhat strange at first, when examining the issue, it becomes apparent that it is not trivial. Teaching memory-impaired patients to use memory aids is compensatory per se, and as it was discussed in the first chapters, acceptance of any type of compensation presupposes insight into the chronic nature of the deficit. Furthermore, it indicates the intention of the patients to reach an optimum for independent living. However, patients might conceptualize their situation differently. For example, they might have "decided" to use their intact hand as much as possible, and this might lead to a loss of function of the impaired hand. If this is the case, only a strong and pertinent input will help them in relearning the use of the impaired hand. Therefore, such patients must be ready to accept a temporary therapeutic regime that has a high likelihood of inducing constraints on their everyday behavior. Such an acceptance depends on insight into the impairment as well as on the belief in the potential to improve in functional independence. Teaching memory-impaired patients to use memory aids does not help them increase insight into their impairment, but rather it is just focused on a technical compensation of the disability. This argument can also be turned the other way around. Confronting patients with tasks to increase functional restitution may lead to the development of internal or external strategies to compensate for their problem. Therefore, restitution-oriented training may be an ideal method to force patients to develop compensatory strategies due to the complexity of the training task structure.

There are two empirical studies which exactly demonstrate that such "side effects" of restitutive training (regarding the development of compensatory efforts) are indeed quite common. das Nair and Lincoln (2012) published a study comparing three groups: one receiving compensational training, one receiving restitutive training for memory impairments, and one serving as a self-help control group. The study was a randomized controlled trial with adequate concealment and with a follow-up after 5 and 7 months. Outcome measures were questionnaires scored by the patients themselves; the assessors of the questionnaires were blinded for the group status. The three groups received 10 sessions with a duration of 1.5 h each. They were also given homework after the end of each session to transfer what was learned during the sessions into their everyday life and to practise the use of memory aids and relaxation strategies. Treatments were based on workbooks and therefore standardized. Participants in both memory intervention programs were taught the use of internal memory strategies and errorless-learning techniques. In addition, those in the compensation group were taught how to use external memory aids. The restitution group engaged in exercises to practise encoding and retrieval, which also included attention retraining exercises. Patients in the restitution group were also taught how to encode specific information using Who, What, Where, When, Why, and How questions. Participants in the self-help group were not taught any memory strategies, but they learned relaxation techniques and ways in which they could cope with their condition. Therefore, the self-help group was aware that they were not treated for their memory impairment, and this indicates a methodological problem for a study which defines a subjective measure as its primary outcome variable.

Three patients from the compensation group dropped out of the study, but evaluation was repeated with a last observation carried forward analysis method. The results showed that the treatment did not result in a statistically significant difference between the groups in the Everyday Memory Questionnaire, which was the main outcome variable of the study. However, the compensatory and the restitution group scored higher for the Internal Memory Aids Questionnaire at the follow-up after 5 and 7 months. The compensatory group also documented an increase in the External Memory Aids Questionnaire after 7 months. At 7 months, the compensation group also showed a better emotional state than the restitution group.

Due to the definition of the outcome variable, the results of the das Nair and Lincoln (2012) study are not easy to interpret. Because the two groups that showed improvement in subjective outcome measures were the ones to receive training and information about memory and memory disturbances after organic brain damage, it seems to be quite logical that these groups were sensitized to their memory impairment. Therefore, they answered the questionnaires differently after the treatment as compared to before the treatment. The finding that this was also the case for the restitutive group shows that confrontation with cognitive problems intuitively leads to the development of strategies (which may be either supportive or obstructive for compensating the deficit). A study by Dirette, Hinojosa, and Carnevale (1999) comparing the effect of restitutive versus compensatory training in patients with visual impairment after brain damage revealed a similar result: After the training, patients from the restitutive group answered that they developed compensatory strategies for solving the training tasks with a similar frequency as the patients who were taught compensatory strategies. Thus, which intervention leads to an acceptance and enduring use of external aids depends on the specific patient and cannot be answered without taking the special circumstances into account.

3. A CLASSIFICATION OF MEMORY AIDS AND TO WHAT KIND OF PATIENTS THEY FIT

The German guideline for diagnosing and treating patients with memory impairment defines that teaching patients to compensate their impairment with external memory aids is a common task, independent of the severity of the impairment (Fig. III.3.1). The same holds for the development of daily routines that decrease the memory load required to succeed in performing the tasks.

A first step for increasing the specificity of the guideline is to classify external memory aids according to their complexity and the specific goal of their practice. The main difference in the complexity of external memory aids should be the factor of whether they are controlled by the patient (him- or herself) or another person. The best examples for these two types of memory aids are NeuroPage (which is controlled by a significant other while the patient just responds to its messages), and an electronic calendar (with a

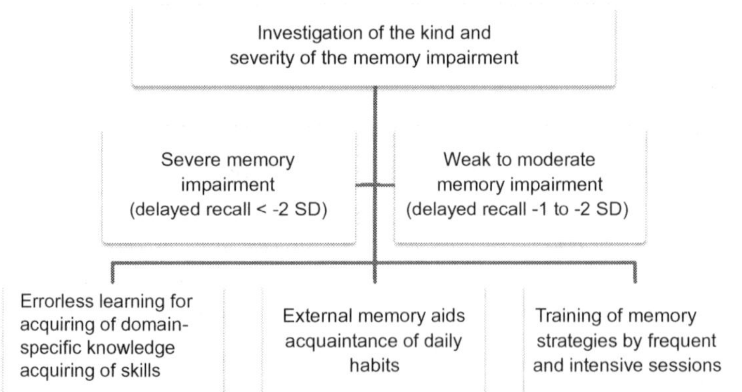

Fig. III.3.1 Flowchart for treating memory impairments from the German Guideline *Diagnostik und Therapie von Gedächtnisstörungen.*

notepad and other submodules which are used by the patients themselves). As reviewed above, electronic calendars may be useful for purely amnesic patients without other cognitive impairments, or by patients with mild memory and cognitive impairments. On the other hand, devices such as NeuroPage are optimal in more-severely impaired patients.

The main difference in goal setting for external memory aids concerns whether they are used to help patients perform a small number of actions (such as taking a drug, calling a person, etc.) or to help them retain some amount of biographical identity. NeuroPage is not suitable to help patients to review the events of the day. To achieve such a goal, external memory aids such as diaries or SenseCams must be used. The distinction according to the level of impairment is important here as well: Severely impaired patients will need help in keeping a diary, whereas mildly impaired patients will be able to do that themselves.

Table III.3.1 provides an overview of which types of memory aids may fit which types of patients. In this table, a key question is whether patients were already acquainted with the use of memory aids before the event that caused the brain damage. The reason for this is that teaching patients with severe memory disorders how to use specific memory aids from the very beginning is a time-consuming therapy and may nevertheless fail. Sohlberg and Mateer (1989) published a single case study that examined the training of an amnesic patient in how to use a compensatory memory book which will be discussed

Table III.3.1 A classification of the usage of external memory aids for different patient groups

<table>
<tr><th></th><th>Severe amnesic disorder plus additional cognitive impairments</th><th colspan="2">Pure amnesic syndrome</th><th>Mild to moderate memory impairment</th></tr>
<tr><td>Patient has used memory aids before the event?</td><td colspan="2">No</td><td colspan="2">Yes</td></tr>
<tr><td>Goal is to teach the patient to execute a specific task?</td><td colspan="2">Pager (NeuroPage, massively simplified smartphones)</td><td colspan="2">Smartphones, tablets, personal digital assistants, calendars and diaries</td></tr>
<tr><td>Goal is to give something like an autobiographical identity?</td><td colspan="2">Passively kept diary, SenseCam</td><td colspan="2">Actively kept diary, SenseCam</td></tr>
</table>

in further detail in the next chapter, because the patients was severely impaired. It took 55 sessions to teach her how to use the memory book, and this was only possible due to the intensive help and involvement of her relatives. Spending such a high number of treatment sessions on one patient in the pursuit of one goal should only be done if the chance to fail with that treatment is low. However, if patients with severe impairments did not use external memory aids before acquiring their impairment, the probability of failure is high (despite intensive and prolonged training). Therefore, starting with such a treatment seems to be risky and cannot be endorsed.

The first row of the table represents the distinction from the German guideline, while the second row indicates whether patients were already acquainted with the use of memory aids (or held knowledge about them) prior to the event that caused their brain damage. The third and fourth row distinguishes between two major goals of the intervention. This can be either to help the patients to achieve an important goal (which has been at the center of most interventional studies), or to help the patients to realize a kind of autobiographical coherence. Just the act of keeping a diary or using a SenseCam can help inform patients of their daily activities (e.g., what he or she has done in the morning as well as events that occurred days or weeks ago). Drawing from clinical knowledge, it is evident that some patients like this kind of objective archiving system.

Moreover, the table is a little bit more explicit through the introduction of the concept of self-use or significant others' use of the external memory aid. The majority of treatment studies analyzing the relation between impairment severity and success in realizing spontaneous use of external memory aids showed that independent use is possible only for patients with no additional cognitive deficits, or only moderate to mild memory impairments. Therefore, in the case of severely impaired patients, a significant other will have to be in charge of updating the memory aid instead of the patients themselves.

We can conclude that for patients with chronic memory disorders who have achieved insight into their impairment, the education on how to use memory aids will be a particularly essential part of their neuropsychological rehabilitation. How patients are instructed to use the memory aid depends on the severity of the memory disorder and the presence of additional cognitive deficits. On the other hand, in patients without clear insight into their impairment and without a high motivation to achieve a higher independence in everyday life, focusing on the introduction of self-used memory aids results in a high rate of failure and passively-used memory-aids should be tried out.

CHAPTER IV

Treatment of Severely Impaired, Amnestic Patients

1. SOME ETIOLOGICAL, EPIDEMIOLOGICAL, AND DEFINING BACKGROUND INFORMATION

Mild to moderate memory impairments are common consequences of brain lesions, while severe memory impairments (which can be called an amnestic syndrome) are rare events and depend on the etiology and the specific localization of the brain lesion. The majority of stroke patients do not suffer from an amnestic syndrome, because most strokes do not affect the central structure of the memory system and involve only one side of the brain. The essential structures for episodic memory (that are well known from clinical and animal studies), are the bilateral Papez circuit, i.e., the hippocampus, the fornix, the mammillary bodies, the anterior nuclei of the thalamus, and the back projections to the hippocampus (indusium griseum). Moreover, the nuclei of the basal forebrain, which are cholinergic and project to the hippocampus and the anterior thalamic nucleus, belong to the core structure of the human memory system. There are multiple diseases where severe amnestic disorders may be expected. Severe traumatic brain injuries, leading to axonal swelling and rupture of fiber structures running in parallel to the acceleration of the trauma, may be such a cause. Herpes encephalitis is another illness that often affects the anterior temporal lobe bilaterally. A bilateral posterior cerebral artery stroke causing neuronal loss of the hippocampus or its afferents can also lead to an amnestic syndrome. Other causes include autoimmune encephalitis targeting relevant receptors of the hippocampus, a global hypoxia due to a cardiac arrest affecting the highly oxygen-dependent neurons of the hippocampus, and rupture of an aneurysm of the anterior communicating artery (which may lead to a loss of the cholinergic nuclei in the basal forebrain). Brain tumors, such as cysts of the third ventricle (and therefore adjacent to the columnae fornicis) or craniopharyngiomas in the direct neighborhood of the mammillary bodies can also cause a full amnestic syndrome.

Cognitive Rehabilitation of Memory
https://doi.org/10.1016/B978-0-12-816981-0.00004-7
© 2019 Elsevier Inc.
All rights reserved.

The neuroanatomy of human memory is beyond the scope and interest of this book. One important aspect of the question of how to treat memory impairments is that they may concern autobiographical memory or the acquisition and access to domain-specific knowledge. Patients with memory impairments may suffer from verbal memory disorders or visuospatial memory impairment. In this case, they do not fulfill the definition of an amnestic syndrome. Broadening this topic, one may distinguish between memory disorders caused by bilateral lesions of the Papez circuit or by a lesion in a region important for domain-specific storage of information: Such a patient is often not amnesic in the classical sense of the word, but, for example, severely impaired in relearning the usage of verbal and written language. These patients "mirror" the situation of a complete amnestic syndrome: They remember what they want to learn but are unable to do so, whereas amnestic patients do not remember what they intend to learn but sometimes know it (e.g., when a patient increases their semantic knowledge of something and cannot remember how they learned it). Both types of severe memory disorders differ in respect to their cognitive architecture (generalized consolidation versus specific storage deficit). However, from a therapeutic point of view, both disorders may call for similar interventions. In this respect, neuropsychological rehabilitation does not differ from other types of rehabilitation: Teaching a patient how to properly use a wheelchair is a common goal, independent of whether the reason is a loss of mobility from an accident, a lesion of the lower spinal cord, or a unilateral cortical stroke affecting the corticospinal tract.

For diagnosing severe memory disorders, the German guideline proposes to use a cut-off score of two standard deviations below the age-, gender-, and education-matched control group. This is a straightforward definition *diagnosing* a severe memory impairment, but not an appropriate cut-off score for developing guidelines for a treatment. Patients with high education levels and young age may score two standard deviations below the average, although their performance still considerably differs from that of a patient with amnesia. For example, a patient with a university education level is allowed to remember seven items in the California Verbal Learning Test (CVLT) in delayed free recall if scoring two standard deviations below the average. Such a free recall performance does not qualify for a diagnosis of amnestic syndrome. In the following chapter(s), a severe memory disorder or an amnestic syndrome will be defined as impairment in delayed free recall, where the patient remembers almost nothing during delayed recall. In terms of the CVLT, the delayed recall score should fall below three words.

The patient should also suffer from some type of orientation disorder. Analogous to motor impairments, this definition also implies a statement on the prospect of rehabilitation: A complete and centrally-caused hemiplegia that exists for more than 4 weeks shows only very little prospect for recovery with the existing methods of treatment. Therefore, a persistent hemiplegia leads to a redefinition of the therapeutic goals towards external compensation.

Patients with (almost) complete amnesia should be tested with easy memory tests (Richter, Thöne-Otto, & Hildebrandt, 2017). For younger patients, one may use the Rivermead Behavioral Memory Test (RBMT), because the subtests of the RBMT are easy and the test battery allows a prognosis for independent living. In older patients, the CERAD-NAB may be used because of its excellent normative basis (at least for German speaking patients). The decision to focus on only a few psychometric tests should be based on the exploration and the anamnesis of the patient. For investigating whether the patient still shows residual memory performance, one may just use four words written on paper. The patient has to remember the words after a short period of intensive distraction. If the patient is unable to do so, an increasing cascade of help should be provided, starting with semantic cues during recall and semantic structuring during an additional time of encoding to phonological and positional cues. If none of these verbal or visuospatial cues allow the patient to remember at least half of the words after having distracted the patient again, one may be sure that the impairment is (almost) complete.

2. TREATMENT OPTIONS

2.1 Errorless Learning and Vanishing Cues in Memory-Impaired Patients

Following the German guidelines, in patients with chronic severe-memory disorders (as defined above) the method of errorless learning should be used to teach specific knowledge, procedures, and daily routines. Specific knowledge to be trained with the method of errorless learning may concern the names of significant others, the routes to be followed to get from "A" to "B", but also may start with training a simple, appropriate behavior in response to a cue (e.g., if an alarm clock is starting to ring). On a theoretical level, one may be skeptical that patients with a complete amnestic syndrome can learn anything, because the definition of their condition is that they are *unable to do so*. On this basis, it should be evident that the learning process used to treat amnestic patients differs from the learning process used in the treatment of patients with mild or moderate memory impairments. In the case of amnestic patients, teaching strategies should not assume that the patients possess a conscious memory trace ("declarative memory") or the internal possibility of an evaluation concerning the correctness of a response. On the contrary: The necessity of a reevaluation of incorrect responses should be avoided. The reason for that is that patients with amnesia often learn incorrect responses as correct (just because the wrong answer becomes more familiar and the patients lack the possibility to suppress the feeling of familiarity by recollection, see the results of Schacter, Verfaellie, Anes, and Racine (1998) and Budson et al. (2002)).

Errorless learning is a method that aims to prevent patients from giving wrong answers. It is a type of shaping and is based on a very frequent repetition of the "question" and the correct "answer". Thus, it is based on implicit or procedural learning, which is often unimpaired or less impaired than explicit learning in amnestic patients. Typically, the patient is confronted with the critical stimulus (e.g., a picture of a significant other) as well as the name of this person, and they are tasked with repeating the written name. The procedure is repeated multiple times ("massed restudy") to ensure that the patient learns the association between picture and name.

One very important (although intensively discussed) method of errorless learning is the method of vanishing cues. Vanishing cues means that the amount of information given to provide the correct answer is reduced over a period of time. A typical example is to learn the name of a person by first presenting the full name, then the full name except the last letter, and then the full name except the two last letters, etc. (Fig. IV.2.1). A reduction of the number of letters only occurs when a correct answer is given. If the answer is correct, the next letter is removed, if it is incorrect, one additional letter is added.

The method of vanishing cues is not restricted to verbal learning. One may also start with the last part of a path and cue the patients to turn in a specific direction until all directional changes have been learned.

Some authors argue that the method of vanishing cues is not part of errorless learning, because reducing the cues may lead to errors. An early review of Kessels and de Haan (2003) even argued that the method of vanishing cues does not help amnesic patients at all. The reason proposed for the absence of any effect of vanishing cues is that the patients are allowed to *guess*, and therefore to perform errors. Indeed, the method of vanishing cues does involve risk that patients perform some error. However, carefully observing

The name of the person in the photograph is …
(please add the missing letters)

Timothy

Timoth

Timot

Timo

Tim

Ti

T

?

The last letter is removed as soon as the patient can write the name correctly. In case of an error, the vanished last letter is added again.

Fig. IV.2.1 An example for the vanishing cues method.

patients for signs of hesitation and intervening in such situations by providing more cues will prevent errors to some degree. Typically, more recent reviews on the method of vanishing cues (Ehlhardt et al., 2008; Ptak, der Linden, & Schnider, 2010) reveal more promising results than the early review of Kessels and de Haan (2003).

Regardless of whether one uses the method of vanishing cues or massed full repetition of a behavior for teaching amnestic patients skills or information, learning with these methods is slow. Furthermore, there is a problem when eliminating the last cue, as the cue is essential to initiate the whole sequence in the mind of the patients. On the other hand, these methods can be extremely effective in teaching specific behavioral performances as long as the context is static. Glisky and Schacter (1989) managed to teach a patient with a dense amnesia the ability to transfer data from forms into data files, although the patient had no experience in using a computer. To achieve that goal, the steps necessary for starting the computer, initiating the program, transferring the important pieces of information, etc., were analyzed into small substeps, which were taught to the patient by the method of vanishing cues. After half a year of training, the patient was able to perform the transfer of data as fast as healthy controls and therefore received a half-time position at the company. The study of Glisky and Schacter (1989) was a single case study. Oudman et al. (2013) showed through a series of single case studies that patients with Korsakoff's syndrome can learn to use a washing machine. Trojano, Moretta, and Estraneo (2009) documented in a single case study that it is possible to teach a young patient suffering from a severe polytrauma to use an eye-tracking communication system. After the treatment, he was able to interact with the environment through that system, but the patient never remembered that he had already used the eye tracker in the past.

In most of the studies, the method of errorless learning was used to teach a specific piece information such as a name of a person or the use of an instrument. Errorless learning and vanishing cues may also be used to teach severely-memory-impaired patients to use an external memory aid, which may help the patients to gain more independence in organizing their everyday lives. Sohlberg and Mateer (1989) demonstrated in a single case study that it is possible to teach the use of an external memory aid (a memory book) to a young patient with anterograde amnesia after herpes encephalitis. Like Glisky and Schacter (1989), they subdivided all steps of using the memory book into substeps and taught the correct execution of these substeps to the patient. By applying the method of vanishing cues, they helped her to transfer these steps into everyday life. Table IV.2.1 shows the different treatment phases including transfer training. Fig. IV.2.2 documents the progress made in teaching the patient to use a memory book in the form of a single

Table IV.2.1 Steps to teach a severely memory-impaired patient to use a memory book (From Sohlberg & Mateer, 1989)

Training phase	Description
Acquisition	Learn names, purpose and use of each notebook section via question/answer format.
Efficiency goal	100% accuracy on questions for five consecutive days.
Application	Learn appropriate methods of recording in notebook via role play situations.
Efficiency goal	100% accuracy of response to three role play situations with no cueing on two consecutive days
Adaptation	Demonstrate appropriate notebook use in naturalistic settings via community training.
Efficiency goal	Receive a score of four for two situations on two consecutive days

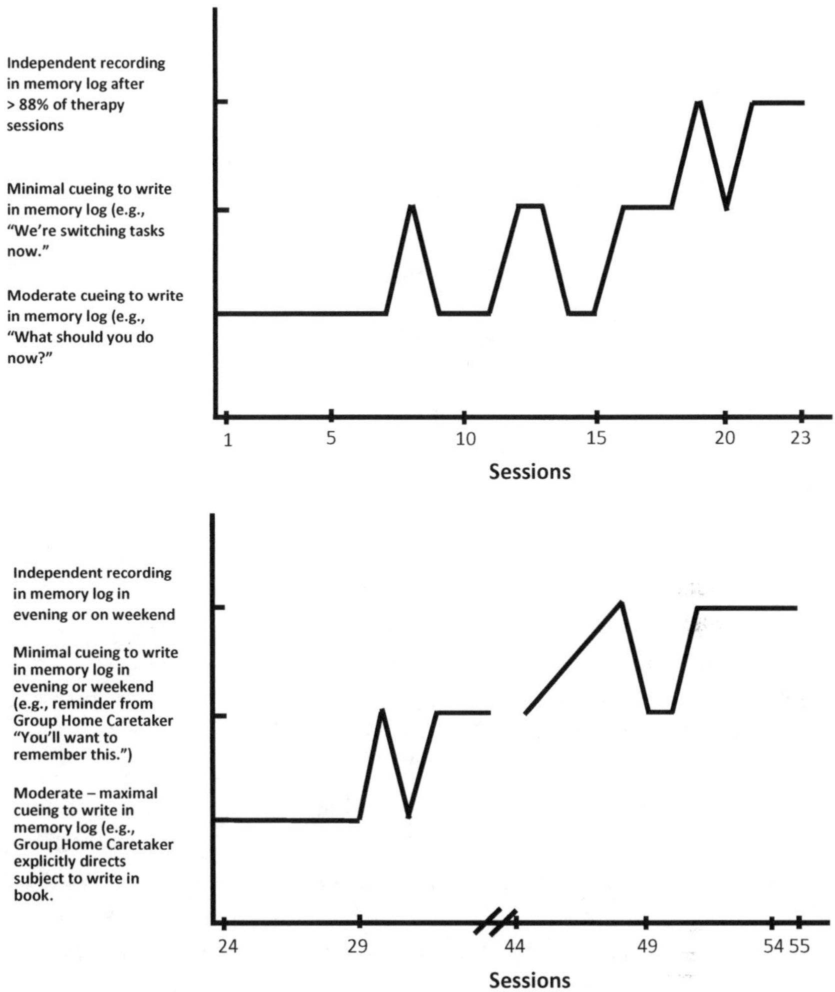

Process graphs depicting level of independence for recording daily events in Memory log. It took 23 days to reach criteria of independent recording of therapy activities during 88 % of the treatment hours. An additional 32 days were required to meet criteria for writing events during evening and weekend hours.

Fig. IV.2.2 Single case experimental design documentation of the progress made in teaching the severely-memory-impaired patient the use of a memory book. *(From Sohlberg M. M., & Mateer C.A. (1989). Training use of compensatory memory books: a three stage behavioral approach.* Journal of Clinical and Experimental Neuropsychology, *11(6), 871–891. https://doi.org/10.1080/01688638908400941 (web archive link).)*

case experimental design plot. The figure also shows that it took more than 50 treatment sessions before the patient could handle the task, despite a high level of engagement of relatives and therapists from other professions in the training.

Pitel et al. (2006) demonstrated in two single case studies that the method of errorless learning can also be helpful in treating patients with combined memory and executive function disorders. It should be mentioned that all these studies relied on single case studies, and it is therefore unclear how many patients might manage with the help of errorless learning to use an external memory aid on their own and how many might not. B. A. Wilson (2013) argued that external memory aids (such as memory books) are the right choice for patients with pure amnesia, as the errorless learning technique allows them to learn their usage fairly reliably. E. de Joode, van Heugten, Verhey, and van Boxtel (2010) and E. A. de Joode, van Heugten, Verhey, and van Boxtel (2013) reviewed publications on external memory aids and attempts to teach memory aids to patients with acquired brain injuries and concluded that increasing the functionality of an external memory aid (as from NeuroPage to a Smartphone, i.e., from a monodirectional to an interactive apparatus) will tend to overwhelm severely-memory-impaired patients. As there is no systematic group study to show the number and types of patients that will benefit from the introduction of an external memory aid, the conclusion of E. de Joode et al. (2010) is important for the topic of this chapter. In general, it is important to state that there are currently no randomized controlled trials (RCTs) involving patients with complete anterograde amnesia as to whether they can learn to use an external memory aid by errorless learning and how far such aids are able to improve independence in everyday life. Therefore, the evidence level for teaching *this* group of patients the independent use of external memory aids by errorless learning is Level IV: It is a clinical consensus which still awaits a RCT. As has been argued before, the evidence level for teaching external memory aids to *moderately to mildly* impaired patients is higher and may be scored as Level IIa. Nevertheless, in the absence of any other possibility to treat severely-memory-impaired patients, one should try to teach them the use of external memory aids.

2.2 Errorless Learning and Vanishing Cues in Patients With Domain-Specific Memory Impairments

As already argued, memory impairment may also concern the ability to relearn domain-specific information. The best example for such an impairment is the relearning of naming and writing after lesions involving the language system of the human brain. Obviously, in these patients the memory system itself is not impaired but rather the specific storage domain for consolidating verbal or graphical information. Errorless learning may be used to retrain the ability to name objects in aphasic patients (as reviewed in Middleton & Schwartz, 2012). One interesting example for treating patients with *severe* aphasia is to teach them to write their own name again. This is a competence of considerable importance, as this would enable these patients to identify themselves if they get lost. In a small and unpublished study, we compared the efficacy of errorless learning with the efficacy of retrieval practice (compare the following chapter) on the ability of severe aphasic patients to write their own names. Fig. IV.2.3 shows the results for two patients.

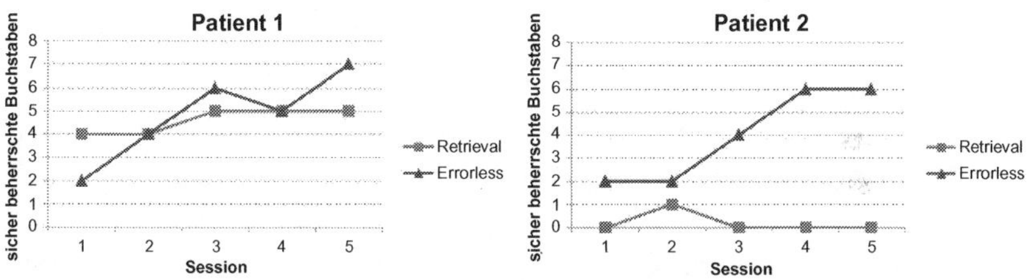

Fig. IV.2.3 Two examples for teaching patients with severe aphasia to write their own name with the methods of errorless learning or retrieval practice.

As one can see, both patients benefitted more from errorless learning than from retrieval practice and learned to write their names in just 1 week of treatment. Middleton and Schwartz (2012), reviewing the comparative efficacy of errorless learning and retrieval practice in the rehabilitation of aphasic patients, argue that errorless learning may be more efficient in short-term but not in long-term outcome. The review of Middleton and Schwartz (2012) is based mainly on studies with mildly and moderately aphasic patients. In a later study, using a crossover design, Middleton, Schwartz, Rawson, and Garvey (2015) and Middleton, Schwartz, Rawson, Traut, and Verkuilen (2016) directly compared errorless learning with retrieval practice in four patients with aphasia to investigate their comparative efficacy. The outcome measure was the performance in object naming. The results showed that errorless learning outperformed retrieval practice directly after the training sessions but not 1 day or 1 week after the training. In these later sessions conducted after the training, the patients treated with the method of retrieval practice achieved a higher object-naming performance. The publication of the study does not include any information about blinding, randomization, concealment, and possible drop-outs, and it is based on just four patients. Therefore, the results are highly interesting from a theoretical point of view, but it is impossible to count it as an RCT. Thus, there are currently no RCTs comparing the methods of errorless learning and retrieval practice with each other or with placebo in patients with aphasia. Therefore, the existing evidence is not higher than Level IIb and more group studies are needed.

Goldenberg and Hagmann (1998) used the method of errorless learning to teach patients with dyspraxia to gain independence in bathing and dressing. Both skills are a major problem for patients with large middle-cerebral artery infarcts and play a role in how long they must stay at an early rehabilitation unit. After only 2–5 weeks of training, most of the patients were able to execute all steps necessary to wash or dress themselves in the morning without performing significant errors. Patients who were forced to stay independent (i.e., they did not receive help from significant others) showed a stable performance during the next 4 weeks, whereas those patients that received help lost their ability. There was no control group for the treatment and the dyspraxia was not really severe. Therefore, this once again necessitates more controlled trials to see whether the combination of errorless learning and self-care training leads to a faster and a more far-reaching relearning in patients with dyspraxia.

During the past few years, a series of studies have been published using errorless learning to teach patients with dementia skills necessary for daily life, food intake, and cognitive tasks, with mixed results (de Werd, Boelen, Rikkert, & Kessels, 2013; Voigt-Radloff et al., 2017).

2.3 Comparing Errorless Learning and Retrieval Practice

Reviewing the treatment approaches for patients with a dense amnesia or domain-specific impairments such as aphasia, three different learning techniques were introduced: errorless learning, vanishing cues, and retrieval practice. Errorless learning and vanishing cues use massed restudy for learning and encoding, but vanishing cues involves stepwise increases of the load on the retrieval process by decreasing the number of cues. Retrieval practice is based on the effect that trying to remember information will increase the stability of the memory trace and therefore the performance in delayed recall. For the rehabilitation of memory-impaired patients, retrieval practice is often combined with the strategy of spacing: The retrieval attempts (almost) directly follow the learning trial and then additional retrieval attempts are performed with an ever-expanding time distance. The combination of retrieval practice with an expanding time lag for the retrieval attempts has been called "spaced retrieval." One may also combine the techniques of errorless learning and vanishing cues with the idea of spacing: In this case, the time between the presentations increases. What is the best method to treat patients with memory impairments? Sumowski et al. (2010) compared errorless learning (massed restudy) with retrieval practice combining the errorless learning with systematic spacing (for a description of the design of the studies refer to Fig. IV.2.4).

Sumowski et al. (2013, 2014) included healthy controls and young patients suffering from a traumatic brain injury (TBI). The inclusion

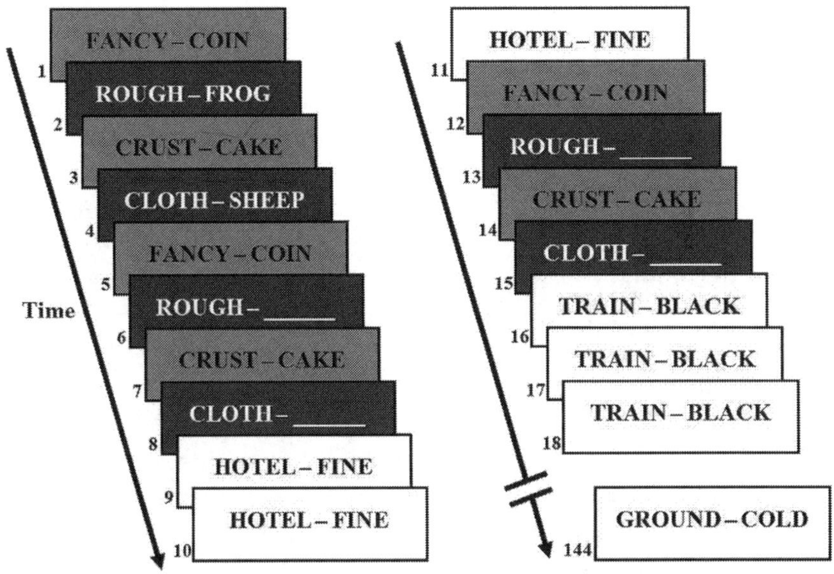

Fig. IV.2.4 Design of the Sumowski, Coyne, Cohen, and Deluca (2014) study. Sample presentation schedule for verbal paired associate (VPAs) in each learning cognition: massed restudy (MR, white), spaced restudy (SR, light gray), retrieval practice (RP, dark gray). Each VPA within the MR condition was presented for three consecutive trials (no intervening trials of other VPAs). Each VPA within the SR or RP condition was presented in a spaced fashion, with three intervening trials (of other VPAs) between the first presentation and first restudy (or test) trial and six intervening trials (of other VPAs) between the first and second restudy (or test) trials. This basic presentation schedule was used for all 48 VPAs, totaling 144 six second trials during a single 14-min 24-s learning phase (6-s slides _ 144 slides) viewed on a computer screen. Note that the 5-s RP trials consisted of a cued-recall screen for 5 s (e.g., CLOTH – ______) followed by a 1-second feedback screen (e.g., CLOTH – SHEEP). Given that VPAs in each condition were evenly distributed throughout the learning phase, any retroactive or proactive interference or practice effects were also equally distributed across groups. Also, VPAs were counterbalanced across the three different conditions across subjects and across delayed.

criterion was a performance in long-delay free recall of the CVLT below one standard deviation of the norm group. The median of the performance of the patients was $z = -2.5$. Spacing of restudy compared to massed restudy was effective in increasing memory performance only in the healthy control group. However, both groups showed the best performance in delayed recall using the retrieval practice for learning the items. In a later study, Sumowski et al. (2014) repeated this comparison for very long-delay recall (after 1 week). They found that TBI patients recalled 11.3% of verbal paired associates when retrieval practice was used as the learning strategy, but they recalled almost nothing at all using massed restudy and spaced restudy. In yet another study (Sumowski et al., 2013), the same effect was demonstrated for patients with multiple sclerosis.

Middleton et al. (2015, 2016) compared retrieval practice and errorless learning (massed restudy) as methods for the rehabilitation of aphasic patients with naming deficits. They also demonstrated that retrieval practice outperformed errorless learning when performance was assessed after a long delay. They mention a reason for the efficacy of both methods: Just repeating the association between a picture and its name means that the name is not retrieved in a top-down fashion from long-term semantic memory; rather, the name is activated from input phonology directly. Therefore, circumventing retrieval from long-term memory may limit the efficacy of errorless learning. Retrieval practice seems to strengthen the top-down search for the name and therefore its *voluntary* use.

These findings give the impression that retrieval practice, but not errorless learning, should be used in memory rehabilitation. This would be especially true for compensatory approaches, where the main task is to teach the patients specific information and not to improve their memory performance. Before rejecting errorless learning as a memory rehabilitation method altogether, it is worthwhile to have a closer look at the patients included in the studies of Sumowski et al. (2010, 2013, 2014) and Middleton et al. (2016). The inclusion criteria of the Sumowski et al. studies were always one standard deviation below the mean of the norm group in delayed recall of the CVLT. For patients with an education level that qualifies them for university entrance, this corresponds to 11 recalled items. The average z-score of the patient group in the Sumowski et al. (2013) TBI study was -2.5. According to the German standardization of the CVLT, that would mean that patients with an educational qualification for university entrance still remembered seven items, and patients that finished secondary school after tenth grade ("Realschulabschluss") remembered six items. In diagnostic terms, achieving a z-score of -2.5 below the age-, gender-, and education-matched comparison group clearly qualifies the performance as severely impaired. As already argued above, in clinical terms such patients do not suffer from a dense amnesia. Patients with a dense amnesia will remember not more than one or two items during the delayed recall condition of the CVLT.

A similar problem arises when inspecting the patients of Middleton et al. (2016). These patients suffered from chronic naming deficits, but not from a global aphasia. Therefore, the results of these studies are interesting for patients with mild to moderate memory impairments, and are discussed in Chapter II under the heading of retrieval based approaches of treatment. However, they cannot be easily generalized to severely impaired patients using a more clinical convention for classifying patients. After all, retrieval practice presupposes that (1) correct retrieval of at least a portion of the learned items is possible after distraction, and that (2) incorrect retrieval can be suppressed by remembering the evaluation of these incorrect answers. These two assumptions do not hold for patients with a dense amnesia or with an *a*phasia but only for those with a *dys*phasia. Therefore, an advantage for retrieval practice compared to errorless learning in these patient groups with severe impairments still must be shown. It seems to be wise to assume that *without a residual memory performance* the method of retrieval practice is not the first choice for treatment. Therefore, a thorough clinical examination of the residual memory performance is essential and must be undertaken before deciding which memory treatment should be used for a specific patient.

3. ORIENTATION TRAINING

It is important to help severely memory-impaired patients regain orientation for time, location, and situation, especially in the early, post-acute phase of rehabilitation. Spector et al. (2003) and Spector, Orrell, and Goyder (2013) were able to show that in patients with progressive dementia, frequent and repeated orientation training increases cognitive performance and quality of life. Unfortunately, for patients with post-acute disorders of orientation, such as patients with severe traumatic brain injury, there are only a few studies that have been conducted (and no study could be classified as an RCT). Reviewing the evidence for orientation training in TBI patients, Langhorn, Sorensen, and Pedersen (2010) argued for its ability to decrease the time of posttraumatic amnesia. A more recent controlled (but not randomized) group study from Langhorn, Holdgaard, Worning, Sorensen, and Pedersen (2015) showed a similar result (decreasing the time of posttraumatic amnesia) and additionally a tendency for a better outcome on the Glasgow Coma Scale. To our knowledge, studies on whether orientation training shows any effect in patients with chronic and severe memory disorders have not been published. In this respect, it would be interesting to know whether a combination of orientation training and the use of a SenseCam (see Chapter II for further information) could improve the results of a simple orientation training. Despite of the lack of evidence-based studies, offering disoriented patients orientation training during early rehabilitation can be claimed as a clinical consensus (Level IV in the EBM hierarchy).

4. TREATING CONFABULATIONS

Problems with memory may primarily concern high rates of forgetting, but it also may involve an extraordinarily high production of memory errors. Memory errors may be explained by a loss of recollection and a high reliance on familiarity. A typical result of such a misbalance is a high number of false-positives during recognition performance. However, memory errors may also result from a loss of post-retrieval monitoring; a frontal selection process that tries to balance remembered information against their context and to decide whether they are relevant for the specific retrieval intention or not. Post-retrieval monitoring has been neuroanatomically related to the ventromedial prefrontal cortex, or even more specifically to the orbitofrontal cortex (Schnider, 2001, 2013). Other authors, reviewing fMRI research, argue that the right mid-ventrolateral prefrontal cortex may play a major role in retrieval monitoring (Nyhus & Badre, 2015). The difference between false-positives in recognition and memory errors due to post-retrieval monitoring errors is that the first are provoked errors whereas the latter may concern spontaneous wrong remembrances. In the latter case, they are called clinically-spontaneous confabulations (Schnider, von Daniken, & Gutbrod, 1996). Confabulations are often combined with severe memory impairments, but this is not always the case: Sometimes the memory performance of these patients is far better than of a patient with a complete amnestic syndrome, and such patients may be free from any type of confabulations (but not from recognition errors). Confabulations pose a considerable problem for neurorehabilitation and for living independently outside of a sheltered environment. They may be defined as assumptions about the situation and about the past which are obviously wrong but are so evident in the minds of the patients that they cannot easily be corrected. Attempts to correct confabulations often lead to serious conflicts with the patients. Confabulations are more frequent during the post-acute phase. Patients who confabulate show self-endangering behavior, because they may try to leave the rehabilitation unit without an orientation to the location of the unit and in a somatic state that does not allow the independence to leave and live on one's own. Such patients often have to be treated in closed units or at psychiatric facilities and may have to take neuroleptic drugs. Unfortunately, these drugs might have a negative impact on the rehabilitation of memory performance and on motor rehabilitation.

During the past few years, several different models have been developed to deliver a cognitive explanation for the development of confabulations (Del Grosso Destreri et al., 2002; Fotopoulou, Conway, & Solms, 2007; Glowinski, Payman, & Frencham, 2008). A generally accepted framework seems to be that at least three different cognitive processes have to be impaired and that additionally the content of the confabulations is often defined by specific emotional responses of the patient towards his or her status. The three cognitive processes are: (1) A disinhibition of a fast and automatic feeling of correctness that is probably produced by the orbitofrontal cortex. This lack of inhibition may also be termed a deficit in post-retrieval monitoring. (2) A disorganized search strategy related to memory contents that are associated with this automatic feeling of correctness (also probably a frontal cortex function). (3) An inability of intellectual reflection on the memory contents. The first deficit, the lack of inhibition, leads to the improper feeling of correctness. The second deficit, the disorganized search, explains the partially fantastic and associative combination of ideas. These often stem from different episodes of the patient's past and lead to a more or less coherent confabulation. The third deficit, the lack of intellectual insight, explains the inability to respond reasonably to arguments raised by the environment against the confabulations. Many confabulating patients tell stories that underline the significance of the patient for a task which must be executed by him or her. Therefore, the content of the confabulation in some respect reflects the threat the patients are experiencing from the situation and the event which led to it, and the attempt to preserve their integrity and autonomy (Fotopoulou, 2008).

There are only a few single case studies on confabulation. Furthermore, the majority do not concern the treatment of this very disabling consequence of a ruptured anterior communicans aneurysm, a frontal basal infarct of the middle cerebral artery, a hypoxia, or a traumatic brain injury. In a recent study, Trivino, Rodenas, Lupianez, and Arnedo (2017) performed a small group study on the treatment of confabulations, which was aimed to reduce the tendency to produce errors in memory tasks. The control group was a waiting-list control group; therefore, the study was not blinded (neither for the patients, nor for the assessors). Interestingly, half of the patients suffered from a large right-hemisphere middle-cerebral artery infarct, a brain lesion rarely reported in single case studies on confabulations. On the other hand, such an infarct exactly concerns brain areas which are associated with post-retrieval monitoring processes. The specific target of the treatment was to increase source memory and the correctness of recency judgments about the remembered information. For the treatment, Trivino et al. (2017) used pictures, words, and biographical memories. Half of the to-be-remembered information was generated by the patient, the other half by the therapist. The patients were instructed to remember whether the information was read or heard and if it was self-generated or presented by the therapist. After a 10-min delay, the information was reproduced by free recall or cued recall. The major topic of the treatment was not so much the performance of memory per se but rather the confrontation of the patients with false reproductions and reducing the feeling of correctness which was present when recalling the false context of the information. To have a significant impact on the overwhelming feeling of correctness experienced by the patients, Trivino et al. (2017) used an ingenious notifying system. This system allowed the patients to control for themselves whether their contextual remembrances were correct or not. After only nine 1-h treatment sessions the number of incorrect source memories was reduced by about 2/3 (Fig. IV.4.1). After 3 months, the reduction was still about 50%, and after an

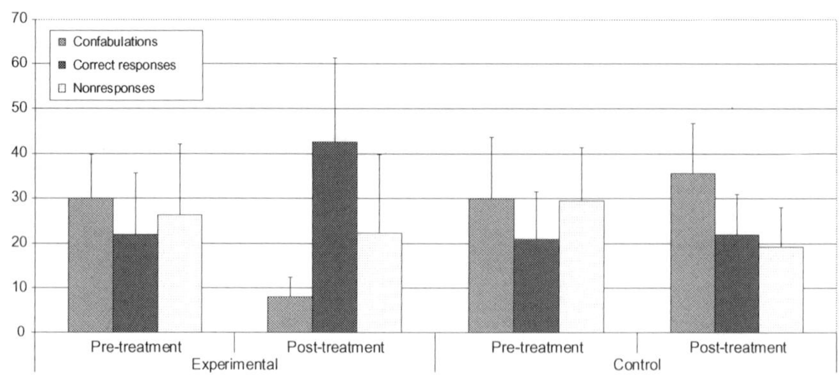

Fig. IV.4.1 Reduction of confabulations in the Trivino et al. (2017) study. Main results for both the experimental (left) and control (right) groups. When we compared the pretreatment and posttreatment measures, the experimental group showed a significant reduction in confabulations and an increase in correct responses, but nonresponses remained the same. However, when the pre-waiting and post-waiting measures of the control group were compared, confabulations did not decrease. In this group nonresponses decreased, while correct responses remained the same.

additional 9 months the patients no longer suffered from confabulations. Because spontaneous recovery from confabulations is common, the final outcome of this study is not as impressive as it might first appear. Nevertheless, the other figures are meaningful enough and it would be very important to have more functional studies on the treatment of confabulations. As confabulations may result in potentially self-endangering consequences for the patients, this is an avenue of rehabilitation research that requires greater exploration.

CHAPTER V

Noncognitive Treatments of Memory Impairments

1. TRANSCRANIAL DIRECT CURRENT STIMULATION

During the past 10 years, transcranial direct current stimulation (tDCS) has been applied in neurological rehabilitation, especially in motor but also in cognitive rehabilitation. The mechanism underlying tDCS is a change in neuronal membrane potentials that is dependent on the direction of the current flow (anodal vs cathodal). Anodal stimulation increases the excitability of the neurons below the electrode and cathodal stimulation decreases the excitability. Therefore, depending on the current flow, it is possible to facilitate or to inhibit neuronal processes. Although the cortical areas that are influenced by tDCS are considerably large and it is impossible to restrict the stimulation to highly specialized areas, some impact on more motor (than, e.g., more sensory or more cognitive) areas can still be realized. One other important limit of tDCS is that the current only reaches cortical surface areas (at least with an effective intensity). Subcortical nuclei or cortical structures not adjacent to the skull cannot be stimulated directly by tDCS. This is of major importance for tDCS as a tool for memory rehabilitation, because recollection is based on hippocampal and thalamic processes. Due to the localization of these two structures they cannot be reached by tDCS. On the other hand, structures like the prefrontal cortex, an area highly associated with encoding and retrieval processes, are open to tDCS stimulation.

Manenti, Cotelli, Robertson, and Miniussi (2012) reviewed tDCS studies on episodic memory performance in healthy subjects and also in patients with brain diseases. They concluded that stimulation of the left dorsolateral prefrontal cortex seems to induce a positive effect on verbal learning and retrieval in healthy controls. A more recent review (Passow, Thurm, & Li, 2017) argues that there is strong evidence for a positive effect of left prefrontal stimulation on verbal memory in older and younger healthy subjects. The effect of temporal lobe stimulation is still unclear, as only a few studies focused on this question. On the other hand, Meier and Sauter (2018) did

Cognitive Rehabilitation of Memory
https://doi.org/10.1016/B978-0-12-816981-0.00005-9
© 2019 Elsevier Inc.
All rights reserved.

not find any effect of tDCS to frontal and parietal brain regions for performance in recognition memory and conclude that memory benefits of tDCS are not easily replicated.

Manenti et al. (2012) reviewed three patient studies using tDCS. The three studies concerned the treatment of Alzheimer's disease patients using anodal tDCS and all showed positive effects. One of the studies used repeated sessions of anodal tDCS applied bilaterally over the temporal cortex. Performance in a visual recognition test was used as the outcome variable. Stimulation with tDCS increased memory performance in this group and the effect was still observed at a follow-up session, 4 weeks after the end of the stimulation. This result is interesting, but because of visual recognition being a rather experimental outcome variable, the transfer to free recall and everyday memory performance remains unclear. Moreover, the study did not include a control group.

Bystad et al. (2016) conducted an RCT with tDCS applied over the left temporal cortex in six 30-min sessions distributed over 10 days. Twenty-five Alzheimer's disease patients were included in the study: 12 received active stimulation and 13 were given sham stimulation. Bystad et al. (2016) used the California Verbal Learning Test (CVLT) as the outcome variable. They found no significant change in the CVLT comparing the two groups, but the increase in delayed recall approached significance ($p = 0.052$). Compared to the study reviewed by Manenti et al. (2012), Bystad et al. (2016) used fewer tDCS sessions and included a very small sample of patients. This and focusing on free recall instead of recognition may explain the differences between the studies. Still, it seems to be fair to argue that evidence for using tDCS in Alzheimer's patient to improve their memory performance is missing. However, there are several trials running on this topic, which will help to clarify the prospect of tDCS in this patients group.

Two studies used tDCS to treat patients with memory impairments resulting from a stroke. Yun, Chun, and Kim (2015) randomized 45 stroke patients to three groups: one receiving anodal stimulation of the left anterior temporal lobe, one receiving stimulation of the right anterior temporal lobe, and one group receiving sham stimulation. Stimulation was delivered five times a week for 30 min, for 3 weeks. The study was double-blinded, but no information is available about concealment. Also, the definition of a primary outcome variable is somewhat vague, as the defined test (a verbal learning test) allowed for different evaluations (learning sum, delayed recall, recognition, etc.). Inclusion criteria were experiencing a stroke within the last 6 months, no lesion concerning the temporal lobes, and a score lower than 27 in the Mini Mental State Examination (MMSE). The exclusion of patients with lesions in the temporal lobe is straightforward because the intervention concerned tDCS of this brain region. On the other hand, such an exclusion criterion implies that patients with hippocampal lesions or lesions in adjacent regions belonging to the limbic system would not be treated. The 15 days of tDCS stimulation led to a significant increase in delayed recall for the verbal learning test, but not group differences in absolute scores. Other tests, such as the MMSE and tests for attention and working memory were not differentially changed by the stimulation. Thus, the impact of the 15 sessions of tDCS was restricted to one neuropsychological outcome variable, did not lead to different performance levels between the groups (but to a higher intragroup increase of performance), and there was no follow-up; two aspects that weaken the significance of this study.

The second study on the impact of tDCS on stroke patients was an experimental study, which only looked for short-term after-effects of stimulation (Kazuta et al., 2017). They also used stimulation of the temporal lobe and found that stimulation increased the learning ability of the stroke patients.

As yet, there have not been any studies combining tDCS and functional memory training or studies on the combination of tDCS and teaching the patients different strategies for encoding and retrieval. Moreover, the patient studies did not rely on left prefrontal stimulation. The current research indicates that stimulating this region shows the most convincing effect in healthy subjects. Because of the important role of the prefrontal cortex in encoding and retrieval (analyzed in Chapter II), one would expect a major impact of its stimulation not only in healthy subjects but also in patients with acquired brain damage who do not suffer from a lesion in the Papez circuit. Unfortunately, we have no data on this question.

In summary, it seems fair to state that tDCS might have an impact on memory rehabilitation, but this still must be shown in "real world" rehabilitation studies.

2. NEUROFEEDBACK

Neurofeedback has been used as a treatment of developmental disorders such as attention deficit hyperactivity disorder (ADHD), but there are only a few studies focusing on the prospect of this method in patients with memory deficits due to acquired brain damage. Recently, Renton, Tibbles, and Topolovec-Vranic (2017) reviewed the application of neurofeedback in stroke patients with memory impairments. The systematic review identified eight studies that could be included into the qualitative synthesis. Five of these studies were case studies, three were group studies, two of the three studies were nonrandomized and none of the group studies were blinded. One of the two group studies that also assessed memory performance showed improved memory performance and one study did not show an improvement. Overall, publications scored poorly concerning both external and internal validity categories, i.e., they showed a high risk of a confounding and selection bias. Power was also insufficient for most of the publications (case studies). Therefore, at this point in time, the significance of neurofeedback for memory rehabilitation is at best unclear.

REFERENCES

Akerlund, E., Esbjornsson, E., Sunnerhagen, K. S., & Bjorkdahl, A. (2013). Can computerized working memory training improve impaired working memory, cognition and psychological health? *Brain Injury*, *27*(13–14), 1649–1657. https://doi.org/10.3109/02699052.2013.830195.

Anderson, N. D., Ebert, P. L., Grady, C. L., & Jennings, J. M. (2018). Repetition lag training eliminates age-related recollection deficits (and gains are maintained after three months) but does not transfer: Implications for the fractionation of recollection. *Psychology and Aging*, *33*(1), 93–108. https://doi.org/10.1037/pag0000214.

Atkinson, R. C., & Shiffrin, R. M. (1968). Human memory: a proposed system and its control processes. *The Psychology of Learning and Motivation*vol. II. (p. xi). Oxford, England: Academic Press 249-xi, 249.

Baddeley, A., & Wilson, B. A. (2002). Prose recall and amnesia: implications for the structure of working memory. *Neuropsychologia*, *40*(10), 1737–1743.

Bahar-Fuchs, A., Clare, L., & Woods, B. (2013). Cognitive training and cognitive rehabilitation for mild to moderate Alzheimer's disease and vascular dementia. *Cochrane Database of Systematic Reviews*. *6*, CD003260. https://doi.org/10.1002/14651858.CD003260.pub2.

Berg, I. J., Koning-Haanstra, M., & Deelman, B. G. (1991). Long-term effects of memory rehabilitation: a controlled study. *Neuropsychological Rehabilitation*, *1*(2), 97–111. https://doi.org/10.1080/09602019108401384.

Bergsma, D. P., & van der Wildt, G. (2010). Visual training of cerebral blindness patients gradually enlarges the visual field. *The British Journal of Ophthalmology*, *94*(1), 88–96. https://doi.org/10.1136/bjo.2008.154336.

Bishara, A. J., & Jacoby, L. L. (2008). Aging, spaced retrieval, and inflexible memory performance. *Psychonomic Bulletin & Review*, *15*(1), 52–57.

Boller, B., Jennings, J. M., Dieudonne, B., Verny, M., & Ergis, A. M. (2012). Recollection training and transfer effects in Alzheimer's disease: effectiveness of the repetition-lag procedure. *Brain and Cognition*, *78*(2), 169–177. https://doi.org/10.1016/j.bandc.2011.10.011.

Borella, E., Carretti, B., Riboldi, F., & De Beni, R. (2010). Working memory training in older adults: evidence of transfer and maintenance effects. *Psychology and Aging*, *25*(4), 767–778. https://doi.org/10.1037/a0020683.

Bowen, A., Hazelton, C., Pollock, A., & Lincoln, N. B. (2013). Cognitive rehabilitation for spatial neglect following stroke. *Cochrane Database of Systematic Reviews*. 7, CD003586. https://doi.org/10.1002/14651858.CD003586.pub3.

Budson, A. E., Sullivan, A. L., Mayer, E., Daffner, K. R., Black, P. M., & Schacter, D. L. (2002). Suppression of false recognition in Alzheimer's disease and in patients with frontal lobe lesions. *Brain*, *125*(Pt 12), 2750–2765.

Bußmann-Mork, B. A., Hildebrandt, H., Gießelmann, H., & Sachsenheimer, W. (2000). Behandlung mittelschwerer sprachlicher Gedächtnisstörungen: ein Vergleich mehrerer Methoden. *Neurologie & Rehabilitation*, *6*, 195–204.

Bystad, M., Gronli, O., Rasmussen, I. D., Gundersen, N., Nordvang, L., Wang-Iversen, H., et al. (2016). Transcranial direct current stimulation as a memory enhancer in patients with Alzheimer's disease: a randomized, placebo-controlled trial. *Alzheimer's Research & Therapy*, *8*(1), 13. https://doi.org/10.1186/s13195-016-0180-3.

Cabrera-Gomez, J. A., Echazabal-Santana, N., Porrero-Martin, P., Valenzuela-Silva, C., Rodriguez, C. A., Fuentes-Suarez, I., et al. (2003). Interferon-alpha2b recombinant improved the cognitive dysfunction in patients with relapsing remitting multiple sclerosis. [Article in Spanish]*Revista de Neurologia, 37*(3), 214–220.

Cashdollar, N., Lavie, N., & Duzel, E. (2013). Alleviating memory impairment through distraction. *The Journal of Neuroscience, 33*(48), 19012–19022. https://doi.org/10.1523/JNEUROSCI.1797-13.2013.

Castro Caldas, A., Levin, J., Djaldetti, R., Rascol, O., Wenning, G., Ferreira, J. J., et al. (2017). Critical appraisal of clinical trials in multiple system atrophy: toward better quality. *Movement Disorders*. https://doi.org/10.1002/mds.27080.

Cermak, L. S., & Butters, N. (1972). The role of interference and encoding in the short-term memory deficits of Korsakoff patients. *Neuropsychologia, 10*(1), 89–95.

Cermak, L. S., Butters, N., & Gerrein, J. (1973). The extent of the verbal encoding ability of Korsakoff patients. *Neuropsychologia, 11*(1), 85–94.

Cermak, L. S., Butters, N., & Moreines, J. (1974). Some analyses of the verbal encoding deficit of alcoholic Korsakoff patients. *Brain and Language, 1*(2), 141–150. https://doi.org/10.1016/0093-934X(74)90030-3.

Chernoff, R. A., Martin, D. J., Schrock, D. A., & Huy, M. P. (2010). Neuropsychological functioning as a predictor of employment activity in a longitudinal study of HIV-infected adults contemplating workforce reentry. *Journal of the International Neuropsychological Society, 16*(1), 38–48. https://doi.org/10.1017/S1355617709990828.

Chiaravalloti, N. D., Ibarretxe-Bilbao, N., DeLuca, J., Rusu, O., Pena, J., Garcia-Gorostiaga, I., et al. (2014). The source of the memory impairment in Parkinson's disease: acquisition versus retrieval. *Movement Disorders, 29*(6), 765–771. https://doi.org/10.1002/mds.25842.

Chiaravalloti, N. D., Moore, N. B., Nikelshpur, O. M., & DeLuca, J. (2013). An RCT to treat learning impairment in multiple sclerosis: the MEMREHAB trial. *Neurology, 81*(24), 2066–2072. https://doi.org/10.1212/01.wnl.0000437295.97946.a8.

Chiaravalloti, N. D., Sandry, J., Moore, N. B., & DeLuca, J. (2016). An RCT to treat learning impairment in traumatic brain injury: the TBI-MEM trial. *Neurorehabilitation and Neural Repair, 30*(6), 539–550. https://doi.org/10.1177/1545968315604395.

Ciaramelli, E., Neri, F., Marini, L., & Braghittoni, D. (2015). Improving memory following prefrontal cortex damage with the PQRST method. *Frontiers in Behavioral Neuroscience. 9*, 211https://doi.org/10.3389/fnbeh.2015.00211.

Cicerone, K. D., Dahlberg, C., Kalmar, K., Langenbahn, D. M., Malec, J. F., Bergquist, T. F., et al. (2000). Evidence-based cognitive rehabilitation: Recommendations for clinical practice. *Archives of Physical Medicine and Rehabilitation, 81*(12), 1596–1615. https://doi.org/10.1053/apmr.2000.19240.

Cicerone, K. D., Dahlberg, C., Malec, J. F., Langenbahn, D. M., Felicetti, T., Kneipp, S., et al. (2005). Evidence-based cognitive rehabilitation: updated review of the literature from 1998 through 2002. *Archives of Physical Medicine and Rehabilitation, 86*(8), 1681–1692. https://doi.org/10.1016/j.apmr.2005.03.024.

Collerton, D., Forster, E., & Packham, D. (2014). An exploratory study of the effectiveness of memory AIDS for older people living in supported accommodation. *Journal of Applied Gerontology, 33*(8), 963–981. https://doi.org/10.1177/0733464812456904.

Corbetta, D., Sirtori, V., Castellini, G., Moja, L., & Gatti, R. (2015). Constraint-induced movement therapy for upper extremities in people with stroke. *Cochrane Database of Systematic Reviews. 10*, CD004433. https://doi.org/10.1002/14651858.CD004433.pub3.

das Nair, R., Cogger, H., Worthington, E., & Lincoln, N. B. (2016). Cognitive rehabilitation for memory deficits after stroke. *Cochrane Database of Systematic Reviews. 9*, CD002293. https://doi.org/10.1002/14651858.CD002293.pub3.

das Nair, R., Cogger, H., Worthington, E., & Lincoln, N. B. (2017). Cognitive rehabilitation for memory deficits after stroke: an updated review. *Stroke, 48*(2), e28–e29. https://doi.org/10.1161/STROKEAHA.116.015377.

das Nair, R., & Lincoln, N. B. (2012). Evaluation of rehabilitation of memory in neurological disabilities (ReMiND): a randomized controlled trial. *Clinical Rehabilitation, 26*(10), 894–903. https://doi.org/10.1177/0269215511435424.

das Nair, R., Martin, K. J., & Lincoln, N. B. (2016). Memory rehabilitation for people with multiple sclerosis. *Cochrane Database of Systematic Reviews*. 3, CD008754. https://doi.org/10.1002/14651858.CD008754.pub3.

de Joode, E., van Heugten, C., Verhey, F., & van Boxtel, M. (2010). Efficacy and usability of assistive technology for patients with cognitive deficits: a systematic review. *Clinical Rehabilitation, 24*(8), 701–714. https://doi.org/10.1177/0269215510367551.

de Joode, E. A., van Heugten, C. M., Verhey, F. R., & van Boxtel, M. P. (2013). Effectiveness of an electronic cognitive aid in patients with acquired brain injury: a multicentre randomised parallel-group study. *Neuropsychological Rehabilitation, 23*(1), 133–156. https://doi.org/10.1080/09602011.2012.726632.

de Werd, M. M., Boelen, D., Rikkert, M. G., & Kessels, R. P. (2013). Errorless learning of everyday tasks in people with dementia. *Clinical Interventions in Aging, 8*, 1177–1190. https://doi.org/10.2147/CIA.S46809.

Del Felice, A., Alderighi, M., Martinato, M., Grisafi, D., Bosco, A., Thompson, P. J., et al. (2017). Memory rehabilitation strategies in nonsurgical temporal lobe epilepsy: a review. *American Journal of Physical Medicine & Rehabilitation*. https://doi.org/10.1097/PHM.0000000000000714.

Del Grosso Destreri, N., Farina, E., Calabrese, E., Pinardi, G., Imbornone, E., & Mariani, C. (2002). Frontal impairment and confabulation after herpes simplex encephalitis: a case report. *Archives of Physical Medicine and Rehabilitation, 83*(3), 423–426.

Delis, D. C., Wetter, S. R., Jacobson, M. W., Peavy, G., Hamilton, J., Gongvatana, A., et al. (2005). Recall discriminability: utility of a new CVLT-II measure in the differential diagnosis of dementia. *Journal of the International Neuropsychological Society, 11*(6), 708–715. https://doi.org/10.1017/S1355617705050812.

della Rocchetta, A. I., & Milner, B. (1993). Strategic search and retrieval inhibition: the role of the frontal lobes. *Neuropsychologia, 31*(6), 503–524.

DeLuca, J., Leavitt, V. M., Chiaravalloti, N., & Wylie, G. (2013). Memory impairment in multiple sclerosis is due to a core deficit in initial learning. *Journal of Neurology, 260*(10), 2491–2496. https://doi.org/10.1007/s00415-013-6990-3.

DeLuca, J., Schultheis, M. T., Madigan, N. K., Christodoulou, C., & Averill, A. (2000). Acquisition versus retrieval deficits in traumatic brain injury: implications for memory rehabilitation. *Archives of Physical Medicine and Rehabilitation, 81*(10), 1327–1333. https://doi.org/10.1053/apmr.2000.9390.

D'Esposito, M., & Postle, B. R. (2015). The cognitive neuroscience of working memory. *Annual Review of Psychology, 66*, 115–142. https://doi.org/10.1146/annurev-psych-010814-015031.

Dewar, B. K., Kapur, N., & Kopelman, M. (2016). Do memory aids help everyday memory? A controlled trial of a Memory Aids Service. *Neuropsychological Rehabilitation*, 1–19. https://doi.org/10.1080/09602011.2016.1189342.

Dewar, M., Della Sala, S., Beschin, N., & Cowan, N. (2010). Profound retroactive interference in anterograde amnesia: what interferes? *Neuropsychology, 24*(3), 357–367. https://doi.org/10.1037/a0018207.

Dewar, M., Garcia, Y. F., Cowan, N., & Della Sala, S. (2009). Delaying interference enhances memory consolidation in amnesic patients. *Neuropsychology, 23*(5), 627–634. https://doi.org/10.1037/a0015568.

Dikmen, S. S., Machamer, J. E., Powell, J. M., & Temkin, N. R. (2003). Outcome 3 to 5 years after moderate to severe traumatic brain injury. *Archives of Physical Medicine and Rehabilitation, 84*(10), 1449–1457.

Dirette, D. K., Hinojosa, J., & Carnevale, G. J. (1999). Comparison of remedial and compensatory interventions for adults with acquired brain injuries. *The Journal of Head Trauma Rehabilitation, 14*(6), 595–601.

Doornhein, K., & de Haan, E. H. F. (1998). Cognitive training for memory deficits in stroke patients. *Neuropsychological Rehabilitation, 8*(4), 393–400. https://doi.org/10.1080/713755579.

Dougall, D., Poole, N., & Agrawal, N. (2015). Pharmacotherapy for chronic cognitive impairment in traumatic brain injury. *Cochrane Database of Systematic Reviews. 12*, CD009221. https://doi.org/10.1002/14651858.CD009221.pub2.

Dubourg, L., Silva, A. R., Fitamen, C., Moulin, C. J., & Souchay, C. (2016). SenseCam: a new tool for memory rehabilitation? *Revue Neurologique (Paris), 172*(12), 735–747. https://doi.org/10.1016/j.neurol.2016.03.009.

Duff, K., Schoenberg, M. R., Scott, J. G., & Adams, R. L. (2005). The relationship between executive functioning and verbal and visual learning and memory. *Archives of Clinical Neuropsychology, 20*(1), 111–122. https://doi.org/10.1016/j.acn.2004.03.003.

Dunning, D. L., & Holmes, J. (2014). Does working memory training promote the use of strategies on untrained working memory tasks? *Memory & Cognition, 42*(6), 854–862. https://doi.org/10.3758/s13421-014-0410-5.

Dusankova, J. B., Kalincik, T., Havrdova, E., & Benedict, R. H. (2012). Cross cultural validation of the minimal assessment of cognitive function in multiple sclerosis (MACFIMS) and the brief international cognitive assessment for multiple sclerosis (BICAMS). *The Clinical Neuropsychologist, 26*(7), 1186 1200. https://doi.org/10.1080/13854046.2012.725101.

Ehlhardt, L. A., Sohlberg, M. M., Kennedy, M., Coelho, C., Ylvisaker, M., Turkstra, L., et al. (2008). Evidence-based practice guidelines for instructing individuals with neurogenic memory impairments: what have we learned in the past 20 years? *Neuropsychological Rehabilitation, 18*(3), 300–342.

Ekuni, R., Vaz, L. J., & Bueno, O. F. A. (2011). Levels of processing: the evolution of a framework. *Psychology & Neuroscience, 4*(3), 333–339. https://doi.org/10.3922/j.psns.2011.3.006.

Emslie, H., Wilson, B. A., Quirk, K., Evans, J. J., & Watson, P. (2007). Using a paging system in the rehabilitation of encephalitic patients. *Neuropsychological Rehabilitation, 17*(4–5), 567–581. https://doi.org/10.1080/09602010701381933.

Evans, J. G., Wilcock, G., & Birks, J. (2004). Evidence-based pharmacotherapy of Alzheimer's disease. *The International Journal of Neuropsychopharmacology*, 7(3), 351–369. https://doi.org/10.1017/S1461145704004444.

Fink, F., Rischkau, E., Butt, M., Klein, J., Eling, P., & Hildebrandt, H. (2010). Efficacy of an executive function intervention programme in MS: a placebo-controlled and pseudo-randomized trial. *Multiple Sclerosis, 16*(9), 1148–1151. https://doi.org/10.1177/1352458510375440.

Fish, J., Manly, T., Emslie, H., Evans, J. J., & Wilson, B. A. (2008). Compensatory strategies for acquired disorders of memory and planning: differential effects of a paging system for patients with brain injury of traumatic versus cerebrovascular aetiology. *Journal of Neurology, Neurosurgery, and Psychiatry, 79*(8), 930–935. https://doi.org/10.1136/jnnp.2007.125203.

Fotopoulou, A. (2008). False selves in neuropsychological rehabilitation: the challenge of confabulation. *Neuropsychological Rehabilitation, 18*(5–6), 541–565. https://doi.org/10.1080/09602010802083545.

Fotopoulou, A., Conway, M. A., & Solms, M. (2007). Confabulation: motivated reality monitoring. *Neuropsychologia, 45*(10), 2180–2190. https://doi.org/10.1016/j.neuropsychologia.2007.03.003.

Glisky, E. L., & Schacter, D. L. (1989). Extending the limits of complex learning in organic amnesia: computer training in a vocational domain. *Neuropsychologia, 27*(1), 107–120.

Glowinski, R., Payman, V., & Frencham, K. (2008). Confabulation: a spontaneous and fantastic review. *The Australian and New Zealand Journal of Psychiatry*, *42*(11), 932–940. https://doi.org/10.1080/00048670802415335.

Goldenberg, G., & Hagmann, S. (1998). Tool use and mechanical problem solving in apraxia. *Neuropsychologia*, *36*(7), 581–589.

Griffiths, S. Y., Yamamoto, A., Boudreau, V. G., Ross, L. K., Kozora, E., & Thornton, A. E. (2005). Memory interference in multiple sclerosis. *Journal of the International Neuropsychological Society*, *11*(6), 737–746. https://doi.org/10.1017/S135561770505085X.

Han, J. W., Oh, K., Yoo, S., Kim, E., Ahn, K. H., Son, Y. J., et al. (2014). Development of the ubiquitous spaced retrieval-based memory advancement and rehabilitation training program. *Psychiatry Investigation*, *11*(1), 52–58. https://doi.org/10.4306/pi.2014.11.1.52.

Han, J. W., Son, K. L., Byun, H. J., Ko, J. W., Kim, K., Hong, J. W., et al. (2017). Efficacy of the ubiquitous spaced retrieval-based memory advancement and rehabilitation training (USMART) program among patients with mild cognitive impairment: a randomized controlled crossover trial. *Alzheimer's Research & Therapy*, *9*(1), 39. https://doi.org/10.1186/s13195-017-0264-8.

Hanks, R. A., Jackson, A. M., & Crisanti, L. K. (2016). Predictive validity of a brief outpatient neuropsychological battery in individuals 1–25 years post traumatic brain injury. *The Clinical Neuropsychologist*, *30*(7), 1074–1086. https://doi.org/10.1080/13854046.2016.1194479.

He, D., Zhang, Y., Dong, S., Wang, D., Gao, X., & Zhou, H. (2013). Pharmacological treatment for memory disorder in multiple sclerosis. *Cochrane Database of Systematic Reviews*. *12*, CD008876. https://doi.org/10.1002/14651858.CD008876.pub3.

Hildebrandt, H. (2017). Was sagt der Tierversuch ueber die Dosis-Wirkungs-Beziehung bei der Rehabilitation von Lern- und Gedaechtnisstoerungen? [What is the animal experiment about the dose-response relationship in the rehabilitation of learning and memory disorders? (PSYNDEXalert)]. *Neurologie & Rehabilitation*, *23*(1), 33–38.

Hildebrandt, H., Bussmann-Mork, B., & Schwendemann, G. (2006). Group therapy for memory impaired patients: a partial remediation is possible. *Journal of Neurology*, *253*(4), 512–519. https://doi.org/10.1007/s00415-006-0013-6.

Hildebrandt, H., Clausing, A., Janssen, H., & Moedden, C. (2007). Rehabilitation leichter bis mittelschwerer Gedaechtnisdefizite - mehr Therapie hilft deutlich mehr, aber welche, wie und warum? [Rehabilitation for minor to moderate memory impairments—the amount of treatment sessions is important, but what type of treatment is more successful and why?]. *Neurologie & Rehabilitation*, *13*(3), 135–145.

Hildebrandt, H., Gehrmann, A., Modden, C., & Eling, P. (2011). Enhancing memory performance after organic brain disease relies on retrieval processes rather than encoding or consolidation. *Journal of Clinical and Experimental Neuropsychology*, *33*(2), 257–270. https://doi.org/10.1080/13803395.2010.511471.

Hildebrandt, H., Lanz, M., Hahn, H. K., Hoffmann, E., Schwarze, B., Schwendemann, G., et al. (2007). Cognitive training in MS: Effects and relation to brain atrophy. *Restorative Neurology and Neuroscience*, *25*(1), 33–43.

Hildebrandt, H., Lehmann, P., & Kastrup, A. (2012). Rehabilitation von Gedächtnisdefiziten nach Hirninfarkt. *Nervenheilkunde*, *31*(6), 440–445.

Hildebrandt, H., Muller, S., Bussmann-Mork, B., Goebel, S., & Eilers, N. (2001). Are some memory deficits unique to lesions of the mammillary bodies? *Journal of Clinical and Experimental Neuropsychology*, *23*(4), 490–501. https://doi.org/10.1076/jcen.23.4.490.1234.

Hildebrandt, H., Zieger, A., Engel, A., & Kraft, A. (1999). Kardiosomatische Kopplung als differentieller Indikator für Komatiefe und Effektivität sensorischer stimulation [Cardiosomatic coupling as a differential indicator for coma depth and effectivity of sensorial stimulation]. *Aktuelle Neurologie, 26*(04), 171–179. https://doi.org/10.1055/s-2007-1017627.

Hrobjartsson, A., & Gotzsche, P. C. (2001). Is the placebo powerless? An analysis of clinical trials comparing placebo with no treatment. *The New England Journal of Medicine, 344*(21), 1594–1602. https://doi.org/10.1056/NEJM200105243442106.

Huang, Y., Lai, X. S., & Tang, A. W. (2007). Comparative study of the specificities of needling acupoints DU20, DU26 and HT7 in intervening vascular dementia in different areas in the brain on the basis of scale assessment and cerebral functional imaging. *Chinese Journal of Integrative Medicine, 13*(2), 103–108. https://doi.org/10.1007/s11655-007-0103-z.

Jennings, J. M., Webster, L. M., Kleykamp, B. A., & Dagenbach, D. (2005). Recollection training and transfer effects in older adults: Successful use of a repetition-lag procedure. *Neuropsychology, Development, and Cognition Section B, Aging, Neuropsychology and Cognition, 12*(3), 278–298. https://doi.org/10.1080/138255890968312.

Karpicke, J. D., & Roediger, H. L., III (2008). The critical importance of retrieval for learning. *Science, 319*(5865), 966–968. https://doi.org/10.1126/science.1152408.

Kaschel, R., Sala, S. D., Cantagallo, A., Fahlböck, A., Laaksonen, R., & Kazen, M. (2002). Imagery mnemonics for the rehabilitation of memory: a randomised group controlled trial. *Neuropsychological Rehabilitation, 12*(2), 127–153. https://doi.org/10.1080/09602010143000211.

Kasten, E., Wust, S., Behrens-Baumann, W., & Sabel, B. A. (1998). Computer-based training for the treatment of partial blindness. *Nature Medicine, 4*(9), 1083–1087. https://doi.org/10.1038/2079.

Kazuta, T., Takeda, K., Osu, R., Tanaka, S., Oishi, A., Kondo, K., et al. (2017). Transcranial direct current stimulation improves audioverbal memory in stroke patients. *American Journal of Physical Medicine & Rehabilitation, 96*(8), 565–571. https://doi.org/10.1097/PHM.0000000000000686.

Kerkhoff, G., Hildebrandt, H., Reinhart, S., Kardinal, M., Dimova, V., & Utz, K. S. (2011). A long-lasting improvement of tactile extinction after galvanic vestibular stimulation: two sham-stimulation controlled case studies. *Neuropsychologia, 49*(2), 186–195. https://doi.org/10.1016/j.neuropsychologia.2010.11.014.

Kerkhoff, G., Munssinger, U., & Meier, E. K. (1994). Neurovisual rehabilitation in cerebral blindness. *Archives of Neurology, 51*(5), 474–481.

Kessels, R. P., & de Haan, E. H. (2003). Implicit learning in memory rehabilitation: a meta-analysis on errorless learning and vanishing cues methods. *Journal of Clinical and Experimental Neuropsychology, 25*(6), 805–814. https://doi.org/10.1076/jcen.25.6.805.16474.

King, A. C., & Dwan, C. (2017). Electronic memory aids for people with dementia experiencing prospective memory loss: a review of empirical studies. *Dementia (London)*. https://doi.org/10.1177/1471301217735180.

Klingberg, T. (2010). Training and plasticity of working memory. *Trends in Cognitive Sciences, 14*(7), 317–324. https://doi.org/10.1016/j.tics.2010.05.002.

Kober, S. E., Schweiger, D., Witte, M., Reichert, J. L., Grieshofer, P., Neuper, C., et al. (2015). Specific effects of EEG based neurofeedback training on memory functions in post-stroke victims. *Journal of Neuroengineering and Rehabilitation. 12*, 107https://doi.org/10.1186/s12984-015-0105-6.

Kolb, B. (1995). *Brain development, plasticity, and behavior*. Lawrence Erlbaum Associates.

Kutschmann, M., Bender, R., Grouven, U., & Berg, G. (2006a). Aspects of sample size determination and power calculation illustrated on examples from rehabilitation research. *Rehabilitation (Stuttg), 45*(6), 377–384. https://doi.org/10.1055/s-2006-940113.

Kutschmann, M., Bender, R., Grouven, U., & Berg, G. (2006b). Aspekte der Fallzahlkalkulation und Powerberechnung anhand von Beispielen aus der rehabilitationswissenschaftlic hen Forschung [Aspects of sample size determination and power calculation illustrated on examples from rehabilitation research]. *Rehabilitation, 45*(06), 377–384. https://doi.org/10.1055/s-2006-940113.

Langhorn, L., Holdgaard, D., Worning, L., Sorensen, J. C., & Pedersen, P. U. (2015). Testing a reality orientation program in patients with traumatic brain injury in a neurointensive care unit. *The Journal of Neuroscience Nursing, 47*(1), E2–10. https://doi.org/10.1097/JNN.0000000000000106.

Langhorn, L., Sorensen, J. C., & Pedersen, P. U. (2010). A critical review of the literature on early rehabilitation of patients with post-traumatic amnesia in acute care. *Journal of Clinical Nursing, 19*(21–22), 2959–2969. https://doi.org/10.1111/j.1365-2702.2010.03330.x.

Lawton, M. P., & Brody, E. M. (1969). Assessment of older people: self-maintaining and instrumental activities of daily living. *Gerontologist, 9*(3), 179–186.

Lee, S. B., Park, C. S., Jeong, J. W., Choe, J. Y., Hwang, Y. J., Park, C. A., et al. (2009). Effects of spaced retrieval training (SRT) on cognitive function in Alzheimer's disease (AD) patients. *Archives of Gerontology and Geriatrics, 49*(2), 289–293. https://doi.org/10.1016/j.archger.2008.10.005.

Li, S. C., Schmiedek, F., Huxhold, O., Rocke, C., Smith, J., & Lindenberger, U. (2008). Working memory plasticity in old age: practice gain, transfer, and maintenance. *Psychology and Aging, 23*(4), 731–742. https://doi.org/10.1037/a0014343.

Linden, M., Hawley, C., Blackwood, B., Evans, J., Anderson, V., & O'Rourke, C. (2016). Technological aids for the rehabilitation of memory and executive functioning in children and adolescents with acquired brain injury. *Cochrane Database of Systematic Reviews*. 7, CD011020. https://doi.org/10.1002/14651858.CD011020.pub2.

Lockwood, C. (2017). Cognitive rehabilitation for memory deficits after stroke: a Cochrane review summary. *International Journal of Nursing Studies, 76*, 131–132. https://doi.org/10.1016/j.ijnurstu.2017.02.011.

Loetscher, T., & Lincoln, N. B. (2013). Cognitive rehabilitation for attention deficits following stroke. *Cochrane Database of Systematic Reviews*. *5*, CD002842. https://doi.org/10.1002/14651858.CD002842.pub2.

Lövden, M., Bodammer, N. C., Kuhn, S., Kaufmann, J., Schutze, H., Tempelmann, C., et al. (2010). Experience-dependent plasticity of white-matter microstructure extends into old age. *Neuropsychologia, 48*(13), 3878–3883. https://doi.org/10.1016/j.neuropsychologia.2010.08.026.

Luft, A. R., Macko, R. F., Forrester, L. W., Villagra, F., Ivey, F., Sorkin, J. D., et al. (2008). Treadmill exercise activates subcortical neural networks and improves walking after stroke: a randomized controlled trial. *Stroke, 39*(12), 3341–3350. https://doi.org/10.1161/STROKEAHA.108.527531.

Lundqvist, A., Grundstrom, K., Samuelsson, K., & Ronnberg, J. (2010). Computerized training of working memory in a group of patients suffering from acquired brain injury. *Brain Injury, 24*(10), 1173–1183. https://doi.org/10.3109/02699052.2010.498007.

Maddox, G. B., & Balota, D. A. (2012). Self control of when and how much to test face-name pairs in a novel spaced retrieval paradigm: an examination of age-related differences. *Neuropsychology, Development, and Cognition Section B, Aging, Neuropsychology and Cognition, 19*(5), 620–643. https://doi.org/10.1080/13825585.2011.640658.

Maddox, G. B., Balota, D. A., Coane, J. H., & Duchek, J. M. (2011). The role of forgetting rate in producing a benefit of expanded over equal spaced retrieval in young and older adults. *Psychology and Aging, 26*(3), 661–670. https://doi.org/10.1037/a0022942.

Magni, L. R., Purgato, M., Gastaldon, C., Papola, D., Furukawa, T. A., Cipriani, A., & Barbui, C. (2013). Fluoxetine versus other types of pharmacotherapy for depression. *Cochrane Database of Systematic Reviews*, 7, CD004185. https://doi.org/10.1002/14651858.CD004185.pub3.

Mahan, S., Rous, R., & Adlam, A. (2017). Systematic review of neuropsychological rehabilitation for prospective memory deficits as a consequence of acquired brain injury. *Journal of the International Neuropsychological Society*, 1–12. https://doi.org/10.1017/S1355617716001065.

Manenti, R., Cotelli, M., Robertson, I. H., & Miniussi, C. (2012). Transcranial brain stimulation studies of episodic memory in young adults, elderly adults and individuals with memory dysfunction: a review. *Brain Stimulation*, *5*(2), 103–109. https://doi.org/10.1016/j.brs.2012.03.004.

Meier, B., & Sauter, P. (2018). Boosting memory by tDCS to frontal or parietal brain regions? A study of the enactment effect shows no effects for immediate and delayed recognition. *Frontiers Psychology*, *9*, 867. https://doi.org/10.3389/fpsyg.2018.00867.

McDonald, A., Haslam, C., Yates, P., Gurr, B., Leeder, G., & Sayers, A. (2011). Google calendar: a new memory aid to compensate for prospective memory deficits following acquired brain injury. *Neuropsychological Rehabilitation*, *21*(6), 784–807. https://doi.org/10.1080/09602011.2011.598405.

McDowall, J. (1979). Effects of encoding instructions and retrieval cuing on recall in Korsakoff patients. *Memory & Cognition*, 7(3), 232–239.

Middleton, E. L., & Schwartz, M. F. (2012). Errorless learning in cognitive rehabilitation: a critical review. *Neuropsychological Rehabilitation*, *22*(2), 138–168. https://doi.org/10.1080/09602011.2011.639619.

Middleton, E. L., Schwartz, M. F., Rawson, K. A., & Garvey, K. (2015). Test-enhanced learning versus errorless learning in aphasia rehabilitation. testing competing psychological principles. *Journal of Experimental Psychology Learning, Memory, and Cognition*, *41*(4), 1253–1261. https://doi.org/10.1037/xlm0000091.

Middleton, E. L., Schwartz, M. F., Rawson, K. A., Traut, H., & Verkuilen, J. (2016). Towards a theory of learning for naming rehabilitation: retrieval practice and spacing effects. *Journal of Speech, Language, and Hearing Research*, *59*(5), 1111–1122. https://doi.org/10.1044/2016_JSLHR-L-15-0303.

Miotto, E. C., Savage, C. R., Evans, J. J., Wilson, B. A., Martin, M. G., Balardin, J. B., et al. (2013). Semantic strategy training increases memory performance and brain activity in patients with prefrontal cortex lesions. *Clinical Neurology and Neurosurgery*, *115*(3), 309–316. https://doi.org/10.1016/j.clineuro.2012.05.024.

Mödden, C., Behrens, M., Damke, I., Eilers, N., Kastrup, A., & Hildebrandt, H. (2012). A randomized controlled trial comparing 2 interventions for visual field loss with standard occupational therapy during inpatient stroke rehabilitation. *Neurorehabilitation and Neural Repair*, *26*(5), 463–469. https://doi.org/10.1177/1545968311425927.

Morris, C. D., Bransford, J. D., & Franks, J. J. (1977). Levels of processing versus transfer appropriate processing. *Journal of Verbal Learning and Verbal Behavior*, *16*(5), 519–533. https://doi.org/10.1016/S0022-5371(77)80016-9.

Morrison, K., Szymkowiak, A., & Gregor, P. (2004). Memojog—An interactive memory aid incorporating mobile based technologies. In Mobile Human-Computer Interaction - Mobilehci 2004, Proceedings (Vol. 3160, pp. 481–485).

Morrow, S. A., Drake, A., Zivadinov, R., Munschauer, F., Weinstock-Guttman, B., & Benedict, R. H. (2010). Predicting loss of employment over three years in multiple sclerosis: clinically meaningful cognitive decline. *The Clinical Neuropsychologist*, *24*(7), 1131–1145. https://doi.org/10.1080/13854046.2010.511272.

Müller, S. V., Hildebrandt, H., & Sachsenheimer, W. (2000). Analysis of the need for clinical neuropsychology in inpatient neurological rehabilitation (phases C and D)—an empirical study of 788 patients. *Rehabilitation (Stuttg)*, *39*(1), 17–25. https://doi.org/10.1055/s-2000-14381.

Nyhus, E., & Badre, D. (2015). Memory retrieval and the functional organization of frontal cortex. In D. R. Addis, M. Barense, & A. Duarte (Eds.), *The Wiley handbook on the cognitive neuroscience of memory*. Chichester, UK: John Wiley.

Oudman, E., Nijboer, T. C., Postma, A., Wijnia, J. W., Kerklaan, S., Lindsen, K., et al. (2013). Acquisition of an instrumental activity of daily living in patients with Korsakoff's syndrome: a comparison of trial and error and errorless learning. *Neuropsychological Rehabilitation, 23*(6), 888–913. https://doi.org/10.1080/09602011.2013.835738.

Passow, S., Thurm, F., & Li, S. C. (2017). Activating developmental reserve capacity via cognitive training or non-invasive brain stimulation: Potentials for promoting fronto-parietal and hippocampal-striatal network functions in old age. *Frontiers in Aging Neuroscience. 9*, 33. https://doi.org/10.3389/fnagi.2017.00033.

Piras, F., Borella, E., Incoccia, C., & Carlesimo, G. A. (2011). Evidence-based practice recommendations for memory rehabilitation. *European Journal of Physical and Rehabilitation Medicine, 47*(1), 149–175.

Pitel, A. L., Beaunieux, H., Lebaron, N., Joyeux, F., Desgranges, B., & Eustache, F. (2006). Two case studies in the application of errorless learning techniques in memory impaired patients with additional executive deficits. *Brain Injury, 20*(10), 1099–1110. https://doi.org/10.1080/02699050600909961.

Ptak, R., der Linden, M. V., & Schnider, A. (2010). Cognitive rehabilitation of episodic memory disorders: from theory to practice. *Frontiers in Human Neuroscience*, 4. https://doi.org/10.3389/fnhum.2010.00057.

Raymer, A. M., Ciampitti, M., Holliway, B., Singletary, F., Blonder, L. X., Ketterson, T., et al. (2007). Semantic-phonologic treatment for noun and verb retrieval impairments in aphasia. *Neuropsychological Rehabilitation, 17*(2), 244–270. https://doi.org/10.1080/09602010600814661.

Reisberg, B., Doody, R., Stoffler, A., Schmitt, F., Ferris, S., Mobius, H. J., et al. (2003). Memantine in moderate-to-severe Alzheimer's disease. *The New England Journal of Medicine, 348*(14), 1333–1341. https://doi.org/10.1056/NEJMoa013128.

Renton, T., Tibbles, A., & Topolovec-Vranic, J. (2017). Neurofeedback as a form of cognitive rehabilitation therapy following stroke: a systematic review. *PLoS One. 12*(5) e0177290. https://doi.org/10.1371/journal.pone.0177290.

Richter, K. M., Modden, C., Eling, P., & Hildebrandt, H. (2015). Working memory training and semantic structuring improves remembering future events, not past events. *Neurorehabilitation and Neural Repair, 29*(1), 33–40. https://doi.org/10.1177/1545968314527352.

Richter, K. M., Modden, C., Eling, P., & Hildebrandt, H. (2018). Improving everyday memory performance after acquired brain injury: an RCT on recollection and working memory training. *Neuropsychology*. https://doi.org/10.1037/neu0000445.

Richter, K. M., Thöne-Otto, A., & Hildebrandt, H. (2017). Diagnostik und Behandlung des schweren amnestischen Syndroms. *NeuroTransmitter, 28*(11), 34–43.

Rohling, M. L., Faust, M. E., Beverly, B., & Demakis, G. (2009). Effectiveness of cognitive rehabilitation following acquired brain injury: a meta-analytic re-examination of Cicerone et al.'s (2000, 2005) systematic reviews. *Neuropsychology, 23*(1), 20–39. https://doi.org/10.1037/a0013659.

Romero, B., & Wenz, M. (2002). Concept and effectiveness of a treatment program for patients with dementia and their relatives. Results from the Bad Aibling Alzheimer Disease Therapy Center. *Zeitschrift für Gerontologie und Geriatrie, 35*(2), 118–128.

Rose, N. S., Craik, F. I., & Buchsbaum, B. R. (2015). Levels of processing in working memory: differential involvement of frontotemporal networks. *Journal of Cognitive Neuroscience, 27*(3), 522–532. https://doi.org/10.1162/jocn_a_00738.

Rugg, M. D., Johnson, J. D., & Uncapher, M. R. (2015). Encoding and retrieval in episodic memory. In D. R. Addis, M. Barense, & A. Duarte (Eds.), *The Wiley handbook on the cognitive neuroscience of memory*. Chichester, UK: John Wiley, pp. 84–107.

Sackett, D. L. (1997). Evidence-based medicine. *Seminars in Perinatology, 21*(1), 3–5.

Sackett, D. L. (1998). Evidence-based medicine. *Spine (Phila Pa 1976), 23*(10), 1085–1086.

Sagar, H. J., Sullivan, E. V., Cooper, J. A., & Jordan, N. (1991). Normal release from proactive interference in untreated patients with Parkinson's disease. *Neuropsychologia, 29* (11), 1033–1044.

Sander, C., Eling, P., Hanken, K., Klein, J., Kastrup, A., & Hildebrandt, H. (2016). The impact of MS-related cognitive fatigue on future brain parenchymal loss and relapse: a 17-month follow-up study. *Frontiers in Neurology*, 7, 155. https://doi.org/10.3389/fneur.2016.00155.

Sander, C., Voelter, H. -U., Schlake, H. -P., Eling, P., & Hildebrand, H. (2017). Diagnostik der Fatigue bei Multipler Sklerose [Assessment of fatigue in multiple sclerosis]. *Aktuelle Neurologie*, *44*(04), 252–259. https://doi.org/10.1055/s-0043-104378.

Schacter, D. L., Rich, S. A., & Stampp, M. S. (1985). Remediation of memory disorders: Experimental evaluation of the spaced-retrieval technique. *Journal of Clinical and Experimental Neuropsychology*, 7(1), 79–96. https://doi.org/10.1080/01688638508401243.

Schacter, D. L., Verfaellie, M., Anes, M. D., & Racine, C. (1998). When true recognition suppresses false recognition: evidence from amnesic patients. *Journal of Cognitive Neuroscience*, *10*(6), 668–679.

Schnider, A. (2001). Spontaneous confabulation, reality monitoring, and the limbic system—a review. *Brain Research Brain Research Reviews*, *36*(2–3), 150–160.

Schnider, A. (2013). Orbitofrontal reality filtering. *Frontiers in Behavioral Neuroscience*. 7, 67. https://doi.org/10.3389/fnbeh.2013.00067.

Schnider, A., von Daniken, C., & Gutbrod, K. (1996). The mechanisms of spontaneous and provoked confabulations. *Brain*, *119*(Pt 4), 1365–1375.

Schulz, K. F., Chalmers, I., Hayes, R. J., & Altman, D. G. (1995). Empirical evidence of bias. Dimensions of methodological quality associated with estimates of treatment effects in controlled trials. *JAMA*, *273*(5), 408–412.

Sheldon, S., Macdonald, R. L., & Schweizer, T. A. (2012). Free recall memory performance after aneurysmal subarachnoid hemorrhage. *Journal of the International Neuropsychological Society*, *18*(2), 334–342. https://doi.org/10.1017/S1355617711001780.

Silva, A. R., Pinho, M. S., Macedo, L., Moulin, C., Caldeira, S., & Firmino, H. (2017). It is not only memory: effects of sensecam on improving well-being in patients with mild Alzheimer disease. *International Psychogeriatrics*, *29*(5), 741–754. https://doi.org/10.1017/S104161021600243X.

Silva, A. R., Pinho, M. S., Macedo, L., & Moulin, C. J. A. (2017). The cognitive effects of wearable cameras in mild Alzheimer disease—an experimental study. *Current Alzheimer Research*, *14*(12), 1270–1282. https://doi.org/10.2174/1567205014666170531083015.

Silva, A. R., Pinho, M. S., Macedo, L., & Moulin, C. J. A. (2018). A critical review of the effects of wearable cameras on memory. *Neuropsychological Rehabilitation*, *28*(1), 117–141. https://doi.org/10.1080/09602011.2015.1128450.

Silva, A. R., Pinho, S., Macedo, L. M., & Moulin, C. J. (2013). Benefits of SenseCam review on neuropsychological test performance. *American Journal of Preventive Medicine*, *44*(3), 302–307. https://doi.org/10.1016/j.amepre.2012.11.005.

Smedslund, G., & Myrhaug, H. T. (2017). *Interventions for visual field defects after stroke: a systematic review*. Oslo, Norway: Knowledge Centre for the Health Services at The Norwegian Institute of Public Health (NIPH). ISBN (digital): 978-82-8082-800-2.

Smith, M. L., Leonard, G., Crane, J., & Milner, B. (1995). The effects of frontal- or temporal-lobe lesions on susceptibility to interference in spatial memory. *Neuropsychologia*, *33*(3), 275–285.

Smith, S. M. (1984). A comparison of two techniques for reducing context-dependent forgetting. *Memory & Cognition*, *12*(5), 477–482.

Snaphaan, L., & de Leeuw, F. E. (2007). Poststroke memory function in nondemented patients: a systematic review on frequency and neuroimaging correlates. *Stroke*, *38*(1), 198–203. https://doi.org/10.1161/01.STR.0000251842.34322.8f.

Sohlberg, M. M., & Mateer, C. A. (1989). Training use of compensatory memory books: a three stage behavioral approach. *Journal of Clinical and Experimental Neuropsychology, 11* (6), 871–891. https://doi.org/10.1080/01688638908400941.

Soveri, A., Antfolk, J., Karlsson, L., Salo, B., & Laine, M. (2017). Working memory training revisited: a multi-level meta-analysis of n-back training studies. *Psychonomic Bulletin & Review*. https://doi.org/10.3758/s13423-016-1217-0.

Soveri, A., Karlsson, E. P. A., Waris, O., Gronholm-Nyman, P., & Laine, M. (2017). Pattern of near transfer effects following working memory training with a dual N-Back task. *Experimental Psychology*, *64*(4), 240–252. https://doi.org/10.1027/1618-3169/a000370.

Spector, A., Orrell, M., & Goyder, J. (2013). A systematic review of staff training interventions to reduce the behavioural and psychological symptoms of dementia. *Ageing Research Reviews*, *12*(1), 354–364. https://doi.org/10.1016/j.arr.2012.06.005.

Spector, A., Thorgrimsen, L., Woods, B., Royan, L., Davies, S., Butterworth, M., et al. (2003). Efficacy of an evidence-based cognitive stimulation therapy programme for people with dementia: randomised controlled trial. *The British Journal of Psychiatry*, *183*, 248–254.

Spencer-Smith, M., & Klingberg, T. (2015). Benefits of a working memory training program for inattention in daily life: a systematic review and meta-analysis. *PLoS One*. *10*(3) e0119522. https://doi.org/10.1371/journal.pone.0119522.

Spreij, L. A., Visser-Meily, J. M., van Heugten, C. M., & Nijboer, T. C. (2014). Novel insights into the rehabilitation of memory post acquired brain injury: a systematic review. *Frontiers in Human Neuroscience*. *8*, 993https://doi.org/10.3389/fnhum.2014.00993.

Stamenova, V., Jennings, J. M., Cook, S. P., Gao, F., Walker, L. A., Smith, A. M., et al. (2016). Repetition-lag memory training is feasible in patients with chronic stroke, including those with memory problems. *Brain Injury*, 1–11. https://doi.org/10.3109/02699052.2016.1147076.

Stamenova, V., Jennings, J. M., Cook, S. P., Walker, L. A., Smith, A. M., & Davidson, P. S. (2014). Training recollection in healthy older adults: clear improvements on the training task, but little evidence of transfer. *Frontiers in Human Neuroscience*. *8*, 898. https://doi.org/10.3389/fnhum.2014.00898.

Sumowski, J. F., Coyne, J., Cohen, A., & Deluca, J. (2014). Retrieval practice improves memory in survivors of severe traumatic brain injury. *Archives of Physical Medicine and Rehabilitation*, *95*(2), 397–400. https://doi.org/10.1016/j.apmr.2013.10.021.

Sumowski, J. F., Leavitt, V. M., Cohen, A., Paxton, J., Chiaravalloti, N. D., & DeLuca, J. (2013). Retrieval practice is a robust memory aid for memory-impaired patients with MS. *Multiple Sclerosis*, *19*(14), 1943–1946. https://doi.org/10.1177/1352458513485980.

Sumowski, J. F., Wood, H. G., Chiaravalloti, N., Wylie, G. R., Lengenfelder, J., & DeLuca, J. (2010). Retrieval practice: a simple strategy for improving memory after traumatic brain injury. *Journal of the International Neuropsychological Society*, *16*(6), 1147–1150. https://doi.org/10.1017/S1355617710001128.

Tate, R. L., Perdices, M., McDonald, S., Togher, L., & Rosenkoetter, U. (2014). The design, conduct and report of single-case research: resources to improve the quality of the neurorehabilitation literature. *Neuropsychological Rehabilitation*, *24*(3–4), 315–331. https://doi.org/10.1080/09602011.2013.875043.

Teasdale, T. W., Emslie, H., Quirk, K., Evans, J., Fish, J., & Wilson, B. A. (2009). Alleviation of carer strain during the use of the NeuroPage device by people with acquired brain injury. *Journal of Neurology, Neurosurgery, and Psychiatry*, *80*(7), 781–783. https://doi.org/10.1136/jnnp.2008.162966.

Thöne-Otto, A. (2012). *Leitlinie "Diagnostik und Therapie von Gedächtnisstörungen"*. https://www.dgn.org/images/red_leitlinien/LL_2012/pdf/ll_94_2012_diagnostik_und_therapie_von_gedchtnisstrungen.pdf.

Trivino, M., Rodenas, E., Lupianez, J., & Arnedo, M. (2017). Effectiveness of a neuropsychological treatment for confabulations after brain injury: a clinical trial with theoretical implications. *PLoS One*. *12*(3)e0173166. https://doi.org/10.1371/journal.pone.0173166.

Trojano, L., Moretta, P., & Estraneo, A. (2009). Communicating using the eyes without remembering it: cognitive rehabilitation in a severely brain-injured patient with amnesia, tetraplegia and anarthria. *Journal of Rehabilitation Medicine*, *41*(5), 393–396. https://doi.org/10.2340/16501977-0344.

Vann, S. D., Tsivilis, D., Denby, C. E., Quamme, J. R., Yonelinas, A. P., Aggleton, J. P., et al. (2009). Impaired recollection but spared familiarity in patients with extended hippocampal system damage revealed by 3 convergent methods. *Proceedings of the National Academy of Sciences of the United States of America*, *106*(13), 5442–5447. https://doi.org/10.1073/pnas.0812097106.

Verhagen, A. P., de Vet, H. C., de Bie, R. A., Kessels, A. G., Boers, M., Bouter, L. M., et al. (1998). The Delphi list: a criteria list for quality assessment of randomized clinical trials for conducting systematic reviews developed by Delphi consensus. *Journal of Clinical Epidemiology*, *51*(12), 1235–1241.

Voigt-Radloff, S., de Werd, M. M. E., Leonhart, R., Boelen, D. H. E., Olde Rikkert, M. G. M., Fliessbach, K., et al. (2017). Structured relearning of activities of daily living in dementia: the randomized controlled REDALI-DEM trial on errorless learning. *Alzheimer's Research & Therapy*, *9*(1), 22. https://doi.org/10.1186/s13195-017-0247-9.

Westerberg, H., Jacobaeus, H., Hirvikoski, T., Clevberger, P., Ostensson, M. L., Bartfai, A., et al. (2007). Computerized working memory training after stroke—a pilot study. *Brain Injury*, *21*(1), 21–29. https://doi.org/10.1080/02699050601148726.

Wilson, B. (1987). Single-case experimental designs in neuropsychological rehabilitation. *Journal of Clinical and Experimental Neuropsychology*, *9*(5), 527–544. https://doi.org/10.1080/01688638708410767.

Wilson, B. A. (1987). *Rehabilitation of memory*. New York, NY: Guilford Press.

Wilson, B. A. (2009). *Memory rehabilitation: integrating theory and practice*. New York, NY: Guilford Press.

Wilson, B. A. (2013). Memory deficits. *Handbook of Clinical Neurology*, *110*, 357–363. https://doi.org/10.1016/B978-0-444-52901-5.00030-7.

Wilson, B. A., Emslie, H., Quirk, K., Evans, J., & Watson, P. (2005). A randomized control trial to evaluate a paging system for people with traumatic brain injury. *Brain Injury*, *19* (11), 891–894.

Wilson, B. A., Emslie, H. C., Quirk, K., & Evans, J. J. (2001). Reducing everyday memory and planning problems by means of a paging system: a randomised control crossover study. *Journal of Neurology, Neurosurgery, and Psychiatry*, *70*(4), 477–482.

Wilson, B. A., Evans, J. J., Emslie, H., & Malinek, V. (1997). Evaluation of NeuroPage: a new memory aid. *Journal of Neurology, Neurosurgery, and Psychiatry*, *63*(1), 113–115.

Wilson, B. A., & Kapur, N. (2009). Compensating for memory deficits with memory aids. In B. A. Wilson (Ed.), *Memory rehabilitation: integrating theory and practice*. New York: The Guilford Press.

Wong, D., Sinclair, K., Seabrook, E., McKay, A., & Ponsford, J. (2017). Smartphones as assistive technology following traumatic brain injury: A preliminary study of what helps and what hinders. *Disability and Rehabilitation*, *39*(23), 2387–2394. https://doi.org/10.1080/09638288.2016.1226434.

Wong, D., Wang, Q. J., Stolwyk, R., & Ponsford, J. (2017). Do smartphones have the potential to support cognition and Independence following stroke? *Brain Impairment*, 1–11. https://doi.org/10.1017/BrImp.2017.10.

Yun, G. J., Chun, M. H., & Kim, B. R. (2015). The effects of Transcranial direct-current stimulation on cognition in stroke patients. *Journal of Stroke*, *17*(3), 354–358. https://doi.org/10.5853/jos.2015.17.3.354.

Zihl, J. (1988). Homonymous hemianopsia and its rehabilitation. *Klinische Monatsblätter für Augenheilkunde*, *192*(5), 555–558. https://doi.org/10.1055/s-2008-1050177.

INDEX

Note: Page numbers followed by *f* indicate figures, *t* indicate tables, and *b* indicate boxes.

N

O

P

R

S

T

V

W

Z

Made in the USA
Monee, IL
13 September 2026

63138366R00120